T0135978

UNIVERSITÄT LEIPZIG

Proceedings of the second international Chinese – German Workshop on

Sustainable Development of Industrial Parks

May 14 – 16, 2007 in Beijing, China

Editors:
Robert HOLLAENDER, Chun-you WU, Ning DUAN

Organised by:
Chinese Research Academy of Environmental Sciences,
Institute of Eco-Planning and Development of the Dalian University of Technology,
Institute for Infrastructure and Resources Management of the University of Leipzig

Bibliografische Information der Deutschen Nationalbibliothek

Die Deutsche Nationalbibliothek verzeichnet diese Publikation in der
Deutschen Nationalbibliografie; detaillierte bibliografische Daten sind
im Internet über http://dnb.d-nb.de abrufbar.

ISBN 978-3-8325-2229-2

Logos Verlag Berlin GmbH
Comeniushof, Gubener Str. 47,
10243 Berlin
Tel.: +49 (0)30 42 85 10 90
Fax: +49 (0)30 42 85 10 92
INTERNET: http://www.logos-verlag.de

Table of content

INTRODUCTION

In May 2007 in Beijing, the Chinese Research Academy of Environmental Sciences, the Institute of Eco-Planning and Development of the Dalian University of Technology and the Institute for Infrastructure and Resources Management of the University of Leipzig held the second International Sino-German Workshop on Sustainable Development of Industrial Parks. After the first workshop on Sustainable Management of Industrial Parks which was jointly organised in 2004 in Leipzig by the Institute of Eco-Planning and Development of the Dalian University of Technology and the Institute for Infrastructure and Resources Management of the University of Leipzig, this second workshop attracted even more participants from several universities and scientific institutions, from European and Chinese companies, from local, regional and central environmental protection authorities and from development agencies. Scholars, practitioners and administrators discussed "General Aspects of Sustainable Development", "Sustainable Operation of Companies in Industrial Parks", and "Sustainable Development of Industrial Parks".

Since more than a decade eco-industrial parks have been regarded as the prototype of eco-industrial development and of so called industrial ecosystems. The co-location of cooperating companies facilitates the integration of material and energy flows, thereby reducing emissions, waste and production costs and, as a result, increasing resource efficencies. The search for synergies relates the research into industrial ecosystems with work on cluster developments and regional economics. In addition, economies of scale and scope may result from company and social networks whether or not material flow integration is involved. Even though the metaphor of industrial ecosystems has become very popular it only insufficiently describes the whole range of these developments. Firstly, the term ecosystem implies a closed loop of material flow which is not the case in industrial ecosystems yet. Most industrial parks concentrate on by-product use for other production lines. The return flow from the society and/or the region remains still a largely unsolved problem. Because industrial parks produce a diversity of goods for local, regional and world markets, the complete integration of waste streams, of the venous material flow into the production lines and subsequently again into the consumption systems cannot be organised on the level of an industrial park, but requires a much broader, and in many cases, even a global perspective. Secondly, the economies of scale and cluster effects that are more and more important for regional developments escape the idea of an industrial ecosystem as a production network based on energy and material flow integration. Therefore, the term industrial symbiosis seems to be more appropriate for describing the advantages of single and multi sector networks of production and service industries operating locally or regionally together.

Departing from the extensively published Kalundborg example in Denmark, where residues and waste heat are utilised in a cascade of different companies and businesses, it has been debated since how the development of industrial symbiosis can be brought about, and how full advantage can be taken from energy and material flow integration. An important string of the scientific discussion centred on the question whether industrial symbiosis can be planed or whether it is a development that may or may not emerge under market conditions. Today, through preferential policies many countries support the co-location of companies at multi sector industrial estates as well as single sector commercial districts for regional economic development.

In Germany, the development of industrial parks has caught only little attention of the academic community. This is due to the fact, that the development of industrial parks is not a priority for German governments. Industrial parks were established in the course of economic restructuring of large state owned or private enterprises, motivated by the economic gains of material flow integration and shared infrastructure costs and, in some regions, also as a vehicle to promote regional economic development. There is no national preferential policy for industrial parks yet. Instead, German environmental law concentrates on strict waste management regulations, energy effiency objectives, and water and air quality protection. This elaborated and dense system of environmental protection provisions indirectly supports the development of sectoral and regional networks, even though single rigid waste management regulations may have also detrimental effects in this regard. Therefore, the present volume contains only few German contributions from science. The majority of German participants were practitioners of companies which, with two exceptions, preferred not to include a written contribution to the proceedings of the workshop.

In China, encouraging eco-industrial parks has been made a national policy. Being awarded the label of eco industrial park is a specific status that may be achieved only after fulfilling particular preconditions for environmental protection and management. Eco-industrial parks are also an issue for extensive research and the present volume shows, that there are multiple efforts to introduce the principles of eco-industrial parks and eco-industrial development into a broad range of applications. Not all of these approaches have reached the same degree of practical relevance yet but all are strongly motivated by practical problems on the ground and may lead to solutions important for the further sustainable development of Chinese regions. In this regard the present volume offers an overview over academic activities in the field of eco-industrial development and eco-industrial parks in China.Please take account of the fact that most contributions were written by non-native speakers and that the editors cannot comment on these independent contributions.

Please take account of the fact that most contributions were written by non-native speakers and that the editors cannot comment on these independent contributions.

Robert Hollaender, for the editorial team,
December 2009, Leipzig

Environmental Problems And Countermeasures During The Industrial Park Development In China

Luo Yi

Deputy Director-General,

Department of Science, Technology and Standards, SEPA

1. Development Course of China's Industrial Parks

1.1 Current development status

An Industrial Park is an industrial production region with the related collective industrial enterprises, consummate infrastructures and more efficient comprehensive service systems. At present, there are 54 national-level economic and technological development zones (ETDZ) and 55 state-level high-tech industrial development zones in China [1]. Over the years, various state-level development zones have accelerated economical development, and the major economic indicators have grown significantly (including GDP, industrial output, industrial added value, taxes, import and export and absorption of foreign investment, etc.).

From 1995 to 2004, the average growth rate per annum of GDP in state-level ETDZs was 9.5%, in 2005, the total amount of GDP in the state-level ETDZs reached up to 660 billion Yuan, which accounted for 4.85% of the total GDP of China. During 1998 and 2005, the industrial added value of 53 high-tech zones increased 6.4 times, reaching 682 billion yuan. In 2005, the state high-tech zone's GDP reached 913.07 billion yuan, accounting for 5% of the total GDP [2].

1.2 The three stages in the development process of Industrial Parks in China

The development process of China's Development Zones has undergone three phases:

The first phase (1984-1991) was a creation and exploration period. In this stage, the development zones had a slow development pace, the economic aggregate was small, and technology transfer efforts were not enough.

The second phase (1992-1998) was a period of rapid growth. In this phase, the use of foreign capital had significantly improved. Economic strength, industrial output and economic benefit had substantially increased; the development zones had become an important regional economic development driver. In 1996, the industrial output of Guangzhou Development Zone accounted for 11.2% of the city's industrial output value and the output of Tianjin Development Zone accounted for 18% of total industrial output of the city. Meanwhile, the national development zones blossom everywhere, almost all provinces and municipalities had established various types of development zones, which formed a "development zone craze."

The third phase (1999 onwards) was characterized by adjustment and stabilization development of industrial parks. Through years of development, some development zones have gotten good economic and social benefits, but some problems have occurred at the same time, including excessive quantities of land exploitated and soil wasted. According to national statistics, until August 2004, there were 6,866 different types of development zones in China, whose total planning area was 38,600 m^2.Within this total, the State Council approved the establishment of 171 zones and the provincial government approved the establishment of 1,094 development zones. The remaining 5,601 development zones were set up by the city, county and township governments and departments at all levels. To stop the out-of-order development, in 2003 the state started to rectify the development zones. To the end of 2004, SEPA total withdrew and merged 4813 development zones, accounting for 70.1% of the original total.

1.3 The "Second Venture" of development zones

After 20 years of hard work, China's development zones have become a new economic growth point and the focus on the utilization of foreign capital has become the most dynamic manufacturing base and export base. However, the construction of development zones also faces many problems and challenges, such as industrial layout and restructuring, the management mechanism innovation and conservation of land resources utilization, resource shortages and environmental pollution problems. Faced with the new situation, many development zones carried out the "second venture" strategic vision.

In 2000, the Tianjin Economic and Technological Development Area (TEDA) began to form a new pattern through which TEDA can "walk on two legs" which are "high tech" and "manufacturing", instead of concentrating on manufacturing enterprises only. In the "11th Five-Year" plan of Shanghai, a strategy was made to phase out the inferior enterprises to upgrade the level of land use. They planned to phase out 3000~4000 inferior enterprises in five years. According to the plan, the industrial output of unit land area will reach 60 billion Yuan /km^2 in 2010. Also, Suzhou New and Hi-tech Industrial Development Zone (SND) plans to introduce more than 50 research and high-tech development centers .

2. Main Environmental Problems

The establishment of industrial parks, to a certain extent, will be favorable to the regional sustainable development and environmental protection in our country. The establishment of industrial parks is to congregate similar enterprises, industries closely tied to the chain enterprises in the park. Establishment of industrial parks will enable the intensive land use and efficiency resource use, and can also support improvement of overall technological progress, so industrial development in industrial parks is characterized of high scientific and technological content, good economic returns, low resource consumption and less environmental pollution.

Most of our industrial parks management are in favour of environmental protection work and have achieved "win-win" aspects of economic development and environmental protection. However, there is also a part of the industrial parks which pay little attention to environmental protection, which lead to a lot of environmental problems.

2.1 Lack of environmental management agencies

By the constraints of industrial park management system, the environmental management agency of industrial parks is inadequate, resulting in insufficient environmental regulatory functions. In some industrial parks, environmental management is carried out by the relevant agencies or independent environmental agencies.

In some industrial parks, the approval authority for the construction project is unclear and the project approval procedure is not strict. Individual industrial park EIA and the "three simultaneous" implementation rates are almost zero. Due to the lack of environmental management personnel and weak supervision, in addition to a large number of enterprises in the parks, the problem of pollution leak investigation and leak management is a frequent phenomenon, which restricts the industrial park development.

In 2006, seven departments, including SEPA, the National Development and Reform Commission, the Ministry of Supervision, launched an action on Rectifying Illegal Pollution Discharge Enterprises to Protect the Health of the Public, during the legislation process, the departments checked more than 1,900 industrial parks, in which there are nearly 30,000 enterprises, and the departments also investigated more than 4,000 irregularity enterprises.

2.2 The lack of environmental planning

Planning is the precursor of construction of industrial parks, and the industrial park environmental planning is as important as the industrial park development planning. Although most of the industrial parks will prepare a development plan, most of them will not make the regional environmental planning and environmental impact assessment, so that the lacks of the quantity and quality analysis of environmental impacts of industrial parks, the assessment of environmental capacity and total amount of pollutant permit emission in industrial parks and selection of the industrial structure and layout are always serious problems in the development of industrial parks in China. Meanwhile, the rationality and feasibility of the scale of industrial parks' environmental infrastructure construction, industrial park ecological protection and ecological construction program and the industrial park dynamic environmental management systems also need further studies.

2.3 More serious environmental pollution

Due to the lack of environmental planning, some of the industrial parks' internal layout is chaotic, cross-contamination happens among different industries. In some industrial parks, the light polluting industrial enterprises are put together with heavy polluting enterprises, which results in cross-contamination among these enterprises. Because of the irrational industrial layout and industrial structure, some industrial park's environmental quality is at a low level. Preferential policy, favorable location and accessibility are the main factors why the enterprises go into the industrial parks, but waste treatment has been considered little in the meantime, resulting in industries such as food, clothing, real estate, chemical industry, pesticides are mixed up together, and a large number of wastewater is discharged out of order, which induced impact on the surrounding environment.

The healthy development of development zones and industrial parks would be a fact only when these problems are solved correctly, or we will have to repeat the way of "treatment after pollution".

3. Suggestions on Environmental Protection Countermeasures for Industrial Parks

3.1 Strengthen planning EIA for industrial parks

The planning EIAs (Environmental Impact Assessments) for industrial parks should be carried out in accordance to the law, and for the development zones and various industrial parks who did not implement EIAs, complementary EIAs should be performed.

The aims of regional planning EIAs are to imply the idea of scientific development to strengthen environmental protection from aspects of planning, pollution source control and industrial layout control, and thus to guide the industrial parks on the road of sustainable development and to prevent the environmental pollution and ecological destruction. To help the industrial parks accomplish planning EIAs, a simplified procedure for EIA at construction project level will be provided to help industrial parks providing sound investment environment for enterprises[3].

Technical Guidelines for Environmental Impact Assessment of Development Area (HJ/T131-2003) [4], which was issued by SEPA in 2003, has stipulated the general principles, work contents and requirements of EIAs for regional development of various development zones. The main tasks in planning EIAs include:

(1) Identifying the regional development activities of the industrial park which can result in significant environmental impacts and possible environmental factors which may constrain the development of the industrial park;

(2) Analyzing and determining the environmental bearing capacity of related environmental media, and proposing reasonable total amount control options for the main pollutants;

(3) Demonstrating environmental protection options, in the aspects of the reasonability of the scales, processes and layouts of integrated pollution treatment facilities, and optimizing the outlets and effluent types of the water and air pollutants; and

(4) Performing comparative analysis and comprehensive argumentation for various planning options of the industrial park, including the siting of the industrial park, functional zoning, industrial structure and layout, development scale, infrastructure construction, and environmental protection facilities, etc., and finally proposing suggestions and countermeasures for the plan of the industrial park.

3.2 Promoting construction of Eco-industrial Parks (EIPs)

EIP is the third generation of industrial parks, following ETDZ (the first generation) and high-tech industrial development zone (the second generation). The construction of EIPs is based on eco-industrial theories and cleaner production requirements, this kind of industrial parks have the advantages of lateral coupling, vertical coupling, regional integration and structural flexibility. The construction of EIPs in China has been regarded as a new industrial development mode to resolve structural and regional pollution problems, to adjust industrial structure and industrial layout, and to construct resource-saving and environment-friendly society.

The SEPA started to promote eco-industrial development and to demonstrate EIP construction in China in 1999, in order to promote regional integrated control for environmental pollution. Currently, SEPA has approved 25 national EIP demonstrations, including 9 sector-specific EIPs (involving steel-making, energy, electrolytic aluminum, aluminum oxide, sugarcane sugar-making, paper making, chemical, mining, coke making and environmental protection, etc.), 15 sector-integrate EIPs (ETDZ and high-tech zones), and 1 venous industry based EIP. The EIP demonstration units have obtained a good many progresses and achievements in developing circular economy and constructing eco-industry, by strengthening education and propaganda, implementing all kinds of projects and formulating supporting policies, in combination with local social and economic development status and the characteristics of industrial structure. The demonstrations have had countrywide influence, bringing along the development of local (at provincial and municipal levels) EIPs.

In 2006, in order to guide the construction of EIPs in China, SEPA issued the three national environmental protection standards, i.e. the Standard for Sector-specific Eco-industrial Parks (HJ/T 273-2006), Standard for Sector-integrate Eco-industrial Parks, (HJ/T 274-2006), and the Standard for Venous Industry Based Eco-industrial Parks (HJ/T 275-2006).

In order to widen and deepen the EIP demonstration in China, SEPA, Ministry of Commerce, and Ministry of Science and Technology decided to jointly promote construction of national demonstrative EIPs. SEPA has recently issued *Circulation on Carrying out Construction of National Demonstrative Eco-Industrial Parks* (SEPA File No. *Huanfa* [2007]51) for this action. According to this circulation, the three ministries will jointly construct national demonstrative EIPs in national-level ETDZs and national-level high-tech industrial development zones. This action should be of historic significance, favorable to the eco-melioration of the national-level ETDZs and high-tech industrial development zones, and to the realization of sustainable development.

3.3 Implement stoppage of approval of EIAs with a whole region

In January 2007, SEPA disclosed a list of illegal construction projects to be punished, and announced to exert stoppage of approval of newly-constructed projects in several regions and group companies which had seriously broken environmental laws. This is so-called "stoppage of approval of EIAs with a whole region and group company". The stoppage policy is regarded as an innovative and powerful measure by experts and the public.

The stoppage policy involving the industrial parks include:

(1) For the areas that have not accomplished the abatement targets stipulated in *"Objective Responsibility Statement for Total Amount Abatement of Pollutants"* in time, the approval of constructed projects in these areas which may increase the pollutant effluent will be stopped temporarily.

(2) For the areas that have serious ecological destruction or have not accomplished the ecological rehabilitation tasks, the approval of constructed projects in these areas which may have significant ecological impacts will be stopped temporarily.

(3) For the areas that have not approved EIAs for constructed projects in accordance with the pre-conditions, procedure and authority level-restriction set by the laws, or have not checked the projects by the laws, or the implementation rate of EIAs and "three simultaneousness" is low, rectification within a prescribed time will be required. If the rectification has not been accomplished in time, the approval authority of constructed projects in these areas will be stopped temporarily.

(4) For the industrial development zones that have exceeded total amount control targets and the resulted environmental quality does meet with the standard requirements, the approval of constructed projects in these zones which may increase the pollutant effluent will be stopped temporarily.

3.4 Strengthen environmental supervision and control in industrial parks

(1) Strengthen environmental management in industrial parks by multiple measures

Environmental protection departments should be set up and professional environmental personnel increased in the industrial parks if possible. It is necessary to introduce a market mechanism into the industrial park management and implement emission trade by monetizing environmental load to promote the corporations to protect environment actively for economic benefits. It is also necessary to improve public environmental protection consciousness to participate in environmental management to solve the conflict between absence of environmental management ability and ever-enlarging environmental management regions.

(2) Restrict access to environmental goods, supervise environmental performance and strengthen the enforcement of laws

Follow-up environmental actions on "Rectifying Illegal Pollution Discharge Enterprises to Protect the Health of the Public" will be carried out. Environmentally illegal activities in construction and management of the industrial parks will be cleaned up. The implementing rate of EIAs and "three-simultaneous system" will be improved. Centralized rectifying of heavy polluting industries (e.g. chemical, metallurgy, printing and dyeing, paper making and leather making) and industrial parks with illegal pollution effluent will be organized. Supervision and control on the major companies will be strengthened, installed with on-line monitoring facilities.

4. Conclusion

From the 1980s, the development of industrial parks of China has gotten through more than 20 years. We have gotten lots of achievements during the process of development and also draw many lessons through the process. Environmental issues are the most important lesson which we must pay our attention to in the future development. To be sure we are on the way of sustainable development; there are many strategies we must put into operation such as conduction of planning EIAs and establishment of environmental agencies. There are still more problems that need further studies in the field of industrial parks and the base theory, together with the construction methods of eco-industrial parks also needs further research.

References

[1] Lu Lijun, Qiu Xiaoling. The Development of Industrial Park in China [M]. Beijing: China Economic Press, 2003.

[2] Cheng Gong, Zhang Qiuyun, Li Qiancheng, Li Yunfeng, Li Shuhui. The Development Strategy about the Industrial Park. Beijing: Social Science Academic Press. 2006.

[3] Wu Chunyou. Resource Efficiency and Eco-planning Management. Tsinghua University Press. 2006.

[4] SEPA. Technical Guidelines for Environmental Impact Assessment of Development Area. Standards Press of China. 2003.

1.

GENERAL ASPECTS OF

SUSTAINABLE DEVELOPMENT

Synergetic Models for Regional Renewable Resources Management

Liu Yan

(PhD Candidate of Dalian University of Technology, Liaoning, 116023)

Abstract: Source separation of bioorganic waste is seen as an essential tool to further develop waste management practice in China. With this approach the related European policy is discussed in the paper, further discussion is made on the Chinese policies on 'reducing green house gases', 'applying circular economy' and 'employing renewable energy'. Also, the paper consideres economic aspects of different waste treatment methods.

Keywords: solid waste treatment, BMWM, renewable resources, synergetic model

1 Introduction

In China in 2003 there were 574 solid waste treatment plants and engineered disposal sites in operation. These facilities have a total capacity of 73 Mt/a (219 000 t/d) corresponding to a treatment rate of 49.7% of total waste collected [1]. The comparison of total amount of waste collection and treatment capacities of waste treatment plants is shown in Fig.1. It is estimated that the treated solid waste represents only about 40% of the total solid waste produced all over China.

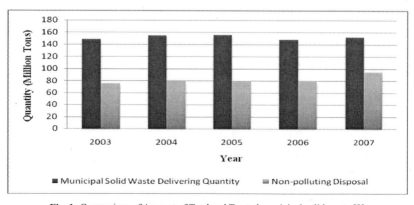

Fig. 1: Comparison of Amount of Total and Treated municipal solid waste [2]

Half of the dumped waste is disposed in 'controlled' landfills, from which a small number (such as in Beijing, Wuhan, Wuxi, Xi'an, Guangzhou) is dealt with in an operating landfill gas recovery system. The Chinese government is facing the rising burden to collect and dispose waste by cost-intensive land filling in increasingly far distance suburban areas. According to municipal solid waste analysis all over China 60% of its content is bioorganic municipal waste [3], which is mainly causing problems during land filling.

About 7% of the collected municipal solid waste is incinerated, (e.g. in Shanghai, Beijing and Shenzhen) but these facilities are working inefficiently due to the high content of organic waste. The incineration plants are fed with low calorific municipal solid waste, and till 2005 there were still more municipal solid waste thermal treatment plants going to be built (such as in Guangzhou, Amoy, Jinjiang, Wenzhou, Huizhou, Kunshan, Puyang, Shijiazhuang and Suzhou). Another 5% of municipal solid waste is treated in mixed waste composting plants [4], and more of such facilities are still going to be built, though the 'output material' can not be used due to quality problems [5].

Chinese authorities have the unanimous opinion that the residents are not able to separate the waste at the source (as in Europe) and they rely on future technologies to separate the waste by technological means. The technologies to separate 'mixed waste' into useable materials do not exist, and attempts in these directions have already failed in the western countries in the 1980s (e.g. Germany, France and Austria). The 'contribution of the general public' by separating their waste at the source, at least in bioorganic 'wet waste' and in 'dry waste', is required. Both fractions can be further treated technologically to achieve material and energy recovery. This collection is the state of the art in all ecologically oriented industrialized countries and is required to implement a cost-effective and economy-relevant waste treatment system, which complies with the objectives of recycling economy [6].

2 Bioorganic municipal waste management in Europe

To tackle the local and global pollution from bioorganic municipal waste disposal Europe has applied a specific bioorganic municipal waste management policy to lessen the negative environmental impact by preventing biodegradable material being disposed of at landfill sites (see EU Directive on Land Filling of Waste, CD 1999/31/EC 1999 and the related national legislations) and based on this the European countries have developed their own 'bioorganic municipal waste strategy' [6,7,8]. Pollution from landfills is mainly caused by bioorganic waste components. One ton of mixed solid waste is producing in total about 300 m³ biogas, containing 55%~65% of CH_4 during the entire landfill duration. From a global point of view, landfill gas (LFG) is one of the anthropogenic sources of green house gases (GHG). According to International Panel on Climate Change (IPCC) 1995 the waste disposal sector is 5.9% of the main sources of GHG emissions.

Between the years from 1800 to 1993 the CH_4 concentrated in the atmosphere had increased from 0.8 to 1.89 ppm and this process is further ongoing. If LFG or biogas is used to substitute fossil energy, renewable CO_2 will replace fossil CO_2. This is the reason why European countries are emphasizing treating bioorganic municipal waste as a priority issue within their waste management policies. The waste management hierarchy of reduction/avoidance, recycling (including posting), treatment (incineration or mechanical biological stabilization) and land filling has been adopted throughout Europe. Governments had achieved the target of 50% recovery rate by 2000, and in 2004 many countries were already far beyond this target. Biological and mechanical biological waste treatment is becoming more and more a key player in meeting national recovery goals, especially as the organic fraction of the household waste stream can be up to 50% or more in some countries. A strong impact derives from the 'EU landfill directive' and its main implications to reduce bioorganic municipal waste by 65% by 2016. Several EU countries have enforced the targets in reducing land filling of bioorganic waste [e.g. >3 %(ω) DM TOC in Germany in July 2005, obligatory, 5 %(ω) DM ignition loss in Austria since 2004 and the collection of BMW on a legal basis in Austria since 1993] far beyond these thresholds. One of the reasons why BMWM within an overall integrated waste management system was successfully developed in the EU was the cost-effectiveness. Biotechnological waste treatment of source-separated waste, even on a high technological standard (including regenerative thermal oxidation, a technology used for flew gas cleaning in Germany when there are high requirements on the output materials) is simply less costly compared to thermal treatment [9].

3 Economic Considerations

In 2002 the National Development and Planning Commission (NDPC), Ministry of Finance (MOF), Ministry of Construction (MOC), and State Environmental Protection Administration (SEPA) released the Announcement No.872, Practicing Charging System of Municipal Solid Waste Treatment and Promoting Industrialization of Waste Treatment [10]. Based on this, some municipalities have started certain attempts to improve waste treatment and to introduce waste disposal charging system models, though the implementation is clearly behind schedule. The waste fees introduced by the communities should be based on the polluter pays principle and full cost covering in order to make the required infrastructure investments sustainable and the economic implications of alternative waste treatment scenarios (incineration versus industrial co-incineration, engineered land filling versus biotechnological bioorganic municipal waste recovery, and their combination within integrated concepts transparent and cost effective). To induce waste avoidance by behavior and participation of the public it is proposed to collect the waste fees from the residential communities. Therefore depending whether a residential communities performs for example well in separation of recyclable materials (packaging and bioorganic municipal waste), and in minimizing mixed waste, the total fee for waste disposal will clearly decrease and the residents will pay less. In this case there is a relation between the behavior of the people living in a residential community and the fees they have to pay for. Waste minimization is stimulated and polluter pays principle is applied (similar to the current system of heating and water fees).

4 Conclusions

Source separation of bioorganic waste is seen as an essential tool to further develop waste management practice in China. With this approach the related European policy is followed, further contribution is made to the Chinese policies on 'reducing green house gases', 'applying circular economy' and 'employing renewable energy'. The ability of the population to carry out source separation of bioorganic municipal waste (a key argument whether going into this direction or not) is in the meantime proven several times under different scenarios, and the public feedback to introduce waste separation is positive. The participants start to mobilize the public media and the local authorities to promote this approach and to extend the pilot areas. Following EU's experience an ongoing visible support of the government over some years is required to adopt such new practice into the daily people's behaviour. Within this campaign, besides of ecological aspects, the positive effect on the future development of waste disposal fees has to be communicated and waste fees systems have to be developed under consideration of the polluter pays principle. Due to the composition of bioorganic municipal waste collected in Chinese cities the combined production of biogas (renewable energy) and 'clean compost' are recommended. In order to improve waste management in urban areas the implementation of a demonstration 'bioorganic municipal waste anaerobic fermentation, composting and thermal utilisation of residual waste plant' should be considered. A plant with a capacity of 100000 t/a can treat the bioorganic municipal waste from about 1 million people. Under consideration of the financial benefits from CDM mechanism (CO_2 emission trading), from electricity (surplus at least between 1.5~2 MW), 5 MW heat for space heating and from compost utilisation, within less than 5 years 50% of the investment costs can be gained. The replacement of fossil CO_2 by using the biogas is related to 700000 trees. The ongoing investigations under the RRU−bioorganic municipal waste project will contribute to build up a solid basis to employ bioorganic municipal waste management in China and it is anticipated to attract more international support to further facilitate this approach and to run full scale pilot projects.

References

[1] Guo Haiqiang, Ma Yingzhen, Zhang Xiaoqin. Status and Problems of City Rubish Processing in China. Journal of Changzhi University, 2007, 24: 104-106.

[2] China Statistical Bureau. China Statistical Yearbook. China Statistical Publishing House, 2004-2008.

[3] Yan Aiping. The Resource Utilization of Municipal Solid Waste in China. Journal of Shengli College China University of Petroleum, 2008, 22(2): 50-52.

[4] Xu H Y. Review on the Development of Municipal Solid Waste Treatment in 2004 [A]. Technology of MSW Treatment in China, No.26 [C]. Beijing: Institute of MSW Management, ESETRC, MOC, 2005. 10−13.

[5] Raninger B. Waste Management Performance in Europe and China [J]. Journal of SY Institute of Aeronautical Engineering, 2002, 19(3): 71−74.

[6] Raninger B, Lorber K, Nelles M, et al. Treatment Strategies for Bioorganic Wastes in Austria [A]. Ministry of Environment Austria BMUJF, Volume 24 [C]. Vienna, Austria, 1999. 1−220.

[7] Raninger B, Bidlingmaier W, Li R D. Management of Municipal Solid Waste in China — Mechanical−Biological reatment can be an Option [A]. Cuvillier Verlag Goettingen, Proceedings of International Symposium Mechanical-biological Waste Treatment (MBT) [C]. Hannover, Germany, 2005. 72−87.

[8] Bidlingmaier W. Compost as a Product [A]. The Necessity of a Real Market (ORBIT) VI. European Forum on Resource and Waste Management [C]. Valencia, Spain, 2002.

[9] Zeng Guanglong. A Simpls Structural Consolidated Exhaust RTO Incinerator. Printed Circuit Information, 2007(5): 20-24.

[10] Biocycle Guide. Anaerobic Digestion [M]. Pennsylvania: JG Press, 2002. 1−78.

Integrated Concepts For Sustainable Industrial Estate Development

Guntram Glasbrenner, Dieter Brulez, Johanna Klein and Andreas König

Abstract

Industrial estate or park planning strategies have changed considerably over the past 30 years to respond to new market demands and changing land-use needs. In Asia, industrial estates have proven a popular means of attracting foreign direct investment by providing full infrastructure and basic services, and often permitting a one-stop service.

In the past, zoning regulations tended to separate industrial uses into categories and have discouraged the mixing of industrial uses with other land uses. However, they are now changing to accommodate new development trends. Locations which offer proximity to a variety of services (e.g. recreation, retail, accommodation and transportation) are becoming highly desirable locations for industrial parks, and no longer purely accommodate traditional industrial developments, separated into specific categories of use.

Industrial parks are increasingly being developed to full urban standards in terms of energy, water, drainage and effluent treatment. Traffic planning and logistics are becoming steadily more important in a global economy with international supply chains and just-in-time production.

Today, modern industrial parks of all types (chemical parks, science and technology parks, supplier parks, commercial parks, etc.) are usually designed to provide a maximum of services in a productive, safe and environmentally sound manner. Industrial parks not only meet environmental and safety standards, but are often at the forefront of setting them for other sectors of industry and commerce. Tailor-made development guidelines provide a higher design standard for new industrial parks. In a modern industrial estate concept, utilities and services often comprise an integral part of the concept and are one of the key factors for investors when it comes to selecting a specific location.

Industrial parks can thus be developed like urban communities, with features not only for production technology but also for social and communal services, commercial and service areas, incubators for business start-ups and other features that benefit each individual company.

This paper presents the relevant activities, methodologies, instruments and achievements of the Deutsche Gesellschaft für Technische Zusammenarbeit (GTZ) GmbH in Asia (including China, India, Indonesia, Malaysia, the Philippines, Thailand, etc.). GTZ is integrating these aspects in its approach to provide an integrated development scenario for the development of new estates and the expansion or redevelopment of existing ones.

1 Theoretical Background

1.1 Integrated industrial estate planning, management and service

Industrial development clearly represents a long-term investment that has to adapt to the multiple aspects of development in a region, as well as to be flexible with regard to the changing economy. It should also create sustained benefits for the economy, and generate and secure employment.

Industrial estate or park planning strategies have changed considerably over the past 30 years in order to respond to new market demands and changing land-use needs. Industrial estates have turned out to be a strong growth motor and have proven a popular means of attracting investment by providing full infrastructure and basic services, and often providing a one-stop service. This is particularly true of the Asian-Pacific region, where they have contributed to the industrialisation process of the newly industrialised countries such as Taiwan, Singapore and South Korea, and have also been one of the decisive factors for stimulating growth in the transitional economies of China and Viet Nam, as well as in newly developed markets such as Thailand and Malaysia. The role of industrial estates as a growth motor for the economy can be especially well illustrated in the Philippines, where export processing zones generate more than 80% of national export revenue. In Europe and the US, the industrial estate approach has become more popular when it comes to upgrading or expanding industries on traditional, single conglomerated sites.

In the past, zoning regulations tended to separate industrial uses into categories such as light, medium or heavy industrial activity, and discouraged the mixing of industrial activity with other land uses. However, this is changing, and nowadays land-use regulations and land-use patterns are increasingly accommodating new development trends. Sites close to a variety of services (e.g. recreation, retail, accommodation and transportation) are becoming highly desirable locations for industrial parks, and these parks are no longer limited to traditional industrial developments separated into specific categories of use.

Industrial parks are increasingly being developed to full urban infrastructure standards in terms of energy, waste and water management, drainage, and effluent treatment. Traffic planning and logistics are becoming ever more important in a global economy with international supply chains and just-in-time production. Selecting the most appropriate traffic mode (rail, road or ship) can, together with innovative traffic management, reduce costs and traffic loads on local roads. This includes connections to the national transportation system, on-site safety, parking facilities, and pedestrian and bicycle circulation.

In the 1970s, industrial park development standards were primarily addressed through zoning by-laws. The implementation of development standards in industrial parks was largely limited, and parks typically suffered from the following negative characteristics:

- limited overall site planning
- limited services and infrastructure
- limited control of design and architectural form and character.

Today, most successful industrial parks (e.g. chemical parks, science and technology parks, supplier parks, commercial parks, etc.) are usually designed to provide a wide range of services in a productive, safe and environmentally sound manner. Industrial parks not only meet environmental and safety standards, but are often at the forefront of setting them for other parts of industry and commerce. Development guidelines ensure that new industrial parks are designed to a higher standard. In modern industrial estate concepts, utilities and services often form an integral part of the concept, and are one of the key factors behind an investor's decision in favour of a specific location.

The master planning process for industrial parks provides the basis for high development standards that are maintained throughout the park. It allows planned environmental settings that have a definable character (e.g. campus-style). Entry and amenity features provide security against theft, and enable entrance controls of trucks and goods. Additionally, they create identity and a feeling of community, and signal to users that they have entered the site. Development standards on adjoining properties can also be consistent with respect to key site elements such as signage, lighting, landscaping, building articulation, parking and fencing.

Industrial parks can thus be developed in much the same way as urban communities, with features not only for production technology but also for social and communal services, commercial and service areas, incubators for business start-ups and other features that benefit individual companies.

GTZ is integrating these aspects into its approach with the aim of providing an integrated scenario for the development of new estates and for the expansion or redevelopment of existing ones. This approach addresses the regional, social and economic setting by developing integrated concepts that benefit local communities and link regional resources. Industrial estates can thus be developed as regional know-how centres with clearly defined benefits for all stakeholders. Such an integrated approach enables the responsible authorities to avoid planning conflicts at an early stage and to recognise stakeholder interests.

1.2 The ECO-Industrial Park Concept

The eco-industrial park (EIP) concept, a new approach for developing industrial estates, was first mentioned during the United Nations Conference on Environment and Development, which took place in Rio in 1992. The concept emerged against the background of increasing worldwide environmental pollution, intensified pressure on natural resources, and growing demand for a more sustainable approach to economic development that causes less negative environmental impact, but at the same time taps into new entrepreneurial benefits by fostering synergy effects.

The objective of developing eco-industrial parks is to ensure the sustainable design and management of industrial estates. The EIP concept can be applied to both existing and new industrial estates. Eco-industrial parks differ from modern industrial parks in two main ways:

• Eco-industrial parks are based on a community of companies working together to share resources and know-how in order to create synergies and benefits that are greater than the sum of any individual benefits.

• The focus of eco-industrial parks includes the whole socio-economic and geographical system at one location, not just the production-oriented aspects within the legal boundaries of an industrial estate. The flow of resources in and out of the estate, impacts on neighbouring areas and development opportunities in the region mean potential synergies for all parties involved.

Modern supply chain systems (e.g. automotive supplier parks) use similar principles, but are designed to achieve single-sector synergies. Eco-industrial parks differ from this in that they integrate available business, design and resource management models to create sustainable industrial communities.

The use of eco-efficiency tools within individual firms, combined with innovative planning and management methods, taps new economic potential through the joint utilisation of infrastructure, production inputs and know-how. Cooperation between companies and local communities on economic, environmental and social issues reduces both conflicts and operating costs, thus creating a win-win situation.

Joint activities in terms of logistics and the use of resources and by-products cut costs, enhance productivity, reduce environmental impacts and make the and make locations more attractive.

To ensure the best possible implementation of an EIP concept, government authorities and private developers should build up a partnership, while at the same time individual companies in industrial estates will be integrated by working on environmental management and cleaner production issues.(for further information, see Lowe, 2001)

2 The GTZ concept for the development of sustainable industrial estates

GTZ has developed an integrated approach for the development of new industrial estates and for the expansion or redevelopment of existing estates.

The GTZ approach is based on improving the political framework conditions, the enhancement of industrial estate planning, management and available services and the shaping of demand for sustainable industrial estates.

2.1 Improving the framework conditions

Working framework conditions are a major prerequisite for smooth operations at a sustainably managed industrial estate. Foreign direct investment will only be attracted to an industrial estate if legal security and compliance with international standards can be guaranteed. GTZ fosters the integration of energy, environmental and social standards into industrial policy and supports the agencies responsible for the management of industrial estates. When it comes to the implementation of a new industrial policy for industrial estates, advice on framework planning and development processes for industrial estates is as important as the networking among state and private sector interest groups.

In many regions, industrial estates are setting their own, more challenging standards than those set by the authorities. These standards are sometimes disseminated after a period of time, enabling these industrial estates to serve as a role model for other industries in the country.

When setting and implementing standards, the management of industrial estates fulfils a dual role: they are setting standards on the one hand, but are also responsible for monitoring and reporting on the other. To be able to fulfil this double role, the industrial estate management needs sufficient staff resources and capacities that are often not provided.

In many countries there is fierce competition for national and international investment between different states and regions. The development of regional investment promotion schemes – tailor-made for a specific region – are an important pillar of the GTZ's activities designed to improve regional competitiveness and increase the attractiveness of an industrial estate to national and international investors.

2.2 Improving industrial estate planning, management and services

2.2.1 Regional master planning, site master planning and assessment of existing sites

Regional Master Planning

Regional master planning (RMP) for industrial regions should aim at balancing the natural, economic and social resources to ensure sustainable economic development. An RMP process has to be in line with the regional and state development goals, and acts as a bridge to the specific town and site master plans.

Sustainable regional development first requires the preparation of a resource assessment to maintain the ecological values of the entire region as a basis for living, agriculture and economic development. Specifically, the protection of water resources for potable supply, agricultural use and ecosystem viability is given high priority in any large-scale industrial development. The protection of ecosystems, species and communities, as well as scenic and historic resources, also has to be included.

Based upon the resource assessment, an analysis is then performed to determine the capacity of the region to accommodate appropriate growth and still sustain critical natural resources. Thus, an RMP will provide the framework to secure the protection of natural resources, while at the same time supporting a sustainable economy. An RMP should define the amount and type of development that an ecosystem can sustain. This will include infrastructure and housing developments as well as alternative developments for settlements and agriculture/fisheries affected by industrial development. In order to integrate environmental, social and economic goals, an RMP process can provide a broader scope, new methods, regional coordination and community participation.

Site Master Planning

To ensure a sustainable design and to make a new industrial estate as attractive as possible, an accurate master plan is of the utmost importance. Many industrial estate agencies or private investors still lack the knowledge to set up an industrial estate tailored to the needs of the market and of international investors. It is only possible to define the strategic focus of an industrial estate if a clear picture has emerged of the regional development options, the attractiveness of the region for investors its market structure and the availability of skilled personnel or investment promotion schemes. Depending on the actual geography and the kind of industries to be located in an industrial estate, the layout, design, technical specifications, logistical requirements and necessary services can be defined.

Based on experience with similar projects, and in view of the possible database of existing concepts, studies and mapping, a three-step approach to site master planning is recommended. This should consist of a conceptual plan (verification of data and information basis, definition of the scope of infrastructure for the target industries and the development of a preliminary master plan), a "pre-basic" plan (i.e. a definition of a master plan that already includes make-or-buy decisions for utilities, waste management and the basic road network, defines permit requirements and necessary safety standards and provides a preliminary cost overview), and an extended master plan (which includes the layout of utility networks, piping and instrumentation diagrams, specifications for underground utilities, pipe racks, utility generation systems, package units, etc., plus the preparation of permit applications and a detailed cost estimate).

In parallel to the setting up of a site master plan, an Environmental Impact Assessment (EIA) also has to be prepared. Tailor-made EIAs require an exact knowledge of local requirements and relevant legislation, as well as experience in developing environmental and socio-economic impact studies.

Assessment of existing industrial estates

One way of assessing existing industrial estates is to use the SMIA ("sustainably manage industrial areas") tool, which has been jointly developed by the United Nations Environment Programme (UNEP), Ecomapping and GTZ. This tool make it easy to assess the most important challenges and problems of an industrial area, and provides support in overcoming the obstacles for a more sustainable and profitable development of an industrial area. The main focus lies in improving the organisational set-up, resource efficiency, corporate image, workplace safety and the profitability of an industrial area.

2.2.2 Management models for industrial estates

There is no "one size fits all" role model on how to manage an industrial estate. GTZ provides assistance and consultancy for different kinds of industrial estate management, e.g. for company networks, industry associations and private sector or state institutions.

Three common management models can be differentiated: public management, public-private partnerships, or private management.

In the case of public management, a very popular form of managing industrial estates in China and other Asian countries is the scenario whereby the municipal government sets up and manages the industrial estate itself.

In a public-private partnership, a governmental administration cooperates with business industry associations.

Finally, there are privately managed industrial estates. In this case, three different options can be differentiated. Either the major user of the park is also responsible for managing the park, or the major investors are the shareholders of the company managing the park , or the park is managed by a separate company.

2.2.3 Enhancing and managing services for industrial estates

Upgrading the service providers or the set-up of a professional service company that provides tailored services to industrial estates creates synergies and boosts the operational effectiveness of complex tasks. Through the joint utilisation of infrastructure, waste disposal, electricity supply, risk management and other services specifically designed for an industrial estate, new economic potential can be tapped and more investment attracted.

Experience shows that most internationally active industrial estates (such as Bayer or Höchst in Germany) are managed by an individual service company created by the park's main shareholders. This service company is responsible for all kinds of services from training, infrastructure and site management, to park safety, organisational development, quality management, and communication with surrounding communities and local authorities.

In Germany this model has proven to be very successful and is also being implemented in some of the largest industrial estates in Asia, such as the Shanghai Chemical Industry Park (SCIP).

2.2.4 Improvement of production processes

An industrial estate can only work as successfully as its resident companies. Fierce competition leads to constant pressure to cut costs, while at the same time increases productivity and enhances the environmental and social performance of producing companies.

For these reasons, consultancy for an industrial park also has to consider the production processes and possible synergies between the individual companies located inside the estate. An optimisation of the cost structure of companies, a more efficient use of raw materials and energy and a reduced non-product output will not only increase the competitiveness of single companies, but will also lead to positive side-effects for the industrial estate. A better information flow and more transparency between the companies due to more innovative management systems provides the possibility to integrate more companies, especially small and medium-sized enterprises (SMEs), into the supply chain. The potential of SMEs (which represent around 90% of overall economic activity in Asia, but often lack the possibility to access finance and knowledge), can be realised by establishing supplier relations inside the industrial park. This strengthens the local economy as well as leads to an increase in overall park productivity and competitiveness.

2.2.5 Integration of local communities into industrial estate activities

Communities located near to an industrial estate do not necessarily benefit from the industrial estate's activities, but instead often suffer from the negative consequences of a high number of industries located in a limited area. Environmental pollution, land-use conflicts and risks can make them major opponents of an industrial estate.

By minimising the concerns of the surrounding communities and cooperating on economic, environmental and social issues it is possible to reduce both conflicts and operating costs, thus creating a win-win situation. Active job creation that improves employment prospects in the region and creates a more accessible workforce for the industrial estate in two main ways.

Firstly, local communities are trained, vocational schools are set up, and workplace centres implemented. Secondly, demand for certain services is boosted, such as restaurants, food stalls, petty trade or modes of transportation. For example, In Thailand GTZ, commissioned by the Industrial Estate Authority of Thailand, supported the Map Ta Phut Industrial Estate in improving its relations with the local community. Due to the poor reputation of the industrial estate, the communities were major opponents concerning expanding its activities. However, this changed with the implementation of an "open door" policy and the establishment of a regional Eco-center, which helped to improve the relations of the communities with the industrial estate. A "Visit Our Neighbour" programme was developed, local communities were trained, and consultancy services were provided for BDS and SME start-ups as well as on environmental topics. Within two years, over 450 projects with a total of 60 companies had presented their projects, which ranged from material savings in terms of training and transportation to joint energy projects and safety plans.

2.3 Shaping demand

Shaping demand for sustainably managed industrial estates is the third area promoted by GTZ.

The international discussion on industrial development suggests that industrial estate concepts should be integrated, especially in the Asian-Pacific region. As quoted by the UN Commission for Sustainable Development (UNCSD), industry has a key role to play in achieving the goals of sustainable development as a supplier of goods and services required by society, as a source of job creation, and as an active participant in community life (UN Commission on Sustainable Development; http://www.un.org/esa/sustdev/sdissues/industry/industry.htm).Industrial estates bundle industrial activity and can thus serve as a regional hub for sustainable development.

One way of promoting sustainable development is to support industries in an industrial estate. An industrial estate that also includes national SMEs in the production chains adds to the value chain and enhances economic growth by increasing exports and employment. Clusters or industrial estates where supporting industries are located close to final production create synergies for all companies. Existing industrial parks have the amount of companies as well as sufficient size to create synergies, but often the interaction between the companies is not developed, making it difficult to get to know companies and their potential in detail. Park management, managers of companies and local authorities have to invest time in improving communication channels. A popular example of the synergies that can be generated in an industrial park is industrial symbiosis at Kalundborg, which demonstrates some of the key elements necessary for successful eco-industrial development. (For further information, see Jacobsen, 2003)

If the companies are arranged along the same value chain, they often already have existing networks and business relationships or common interests that foster community development. Examples of this approach include the large chemical industry plants in Germany, as well as automotive supplier parks.

A possible option to increase the demand for sustainably managed industrial estates still further would be to set up an internationally recognised label for eco-industrial estates. China, with its State Environmental Protection Agency (SEPA) guidelines, which provide initial definitions and procedures for eco-industrial estates and their development for the Chinese administration and industry, is using this approach, which has revealed that there is great demand from industrial estates themselves to obtain this eco-label. In the case of the Yanghzou Eco-industrial Estate, GTZ supported the compilation of the Estate's Master Plan. The plan was approved, and SEPA will now grant the industrial estate the status of a National Eco-industry Demonstration park.

Of course, it is not the role of GTZ to define binding and internationally acceptable minimum standards that should ideally be achieved by participating industries in order for them justifiably to claim to be an eco-industrial estate. However, due to GTZ's widespread experience in this area, it would nevertheless be a challenge to start the discussion on a set of indications that should be used when benchmarking the performance of industrial estates, as this would provide valuable input to the discussion on standards.

3 Future developments and recommendations

GTZ is currently working on industrial estates and clusters in various different countries such as China, the Philippines, Indonesia, Thailand and India.

The GTZ approach differs depending on the type of industries, the size and the management system of the industrial estates, their institutional set-up, etc. For the consultancy of industrial estates, there is no "one size fits all" solution. Setting up and managing an industrial estate are such complex tasks that an individual solution has to be found for every different scenario. At the moment, GTZ is mainly focusing on the further development of two different solutions: consultancy services for industrial clusters, and setting up and managing large-scale industrial estates.

For the consultancy of large industrial estates, GTZ can rely on the one hand on its own experience as well as on a network of partners with international operations, such as Bayer. GTZ has developed an approach that mixes its multistakeholder approach and its worldwide experience in managing complex projects with longstanding expertise regarding industrial estate development in Germany and other European countries; this approach can be adapted to all large-scale industrial estate projects. This concept is based on activities GTZ is currently implementing in India, where GTZ has been commissioned by the Andhra Pradesh Industrial Infrastructure Cooperation to submit an offer for the assessment, site master planning and development of one of the largest industrial sites in India.

However, to ensure balanced future development in Asian countries, concepts should not only rely on the development of large, international industrial estates. To strengthen regional competitiveness, it is equally important to provide concepts for national investors and SMEs. In many regions in Asia, so-called clusters exist that consist mainly of SMEs plus maybe a few larger companies. These industrial clusters are hardly managed as such, yet nevertheless constitute an important pillar of regional economic activity. However, their productivity and competitiveness is sub-optimal due to a lack of knowledge about modern management techniques and limited access to knowledge about such management methods. Additionally, many of these clusters consist of highly polluting firms which neither consider the consequences of their production activities for the local communities surrounding the industrial estate nor the regional environment. To develop fully the potential of industrial clusters, innovative approaches that provide easy-to-implement methods and tools are needed. GTZ, through the EU-Asia Pro Eco grant line, is now implementing a project on the development of industrial clusters in central Java, Indonesia. Here, new and innovative approaches are being implemented with the aim of further developing industrial activity and promoting regional competitiveness, as well as improving environmental performance.

REFERENCES

König, A. W. (2005): "EU-China Environmental Management Cooperation Programme – Industry Development: The Eco-industrial Park Development, A Guide for Chinese Government Officials and Industrial Park Managers".

Jacobsen, N. B. (2003): "The Industrial Symbiosis in Kalundborg, Denmark: An Approach to Cleaner Industrial Production", in: E. Cohen-Rosenthal (ed.), *Eco-Industrial Strategies: Unleashing Synergy between Economic Development and the Environment*, Sheffield: Greenleaf Publishing.

Lowe, E. A. (2001): Eco-industrial Park Handbook for Asian Developing Countries. A Report to the Asian Development Bank, Environment Department, Oakland: RPP International.

UN Commission on Sustainable Development: links to documents and preparations of the CSD on the topic of industrial development are available from: http://www.un.org/esa/sustdev/sdissues/industry/industry.htm

General Tendency in Per Capita Material Consumption and Intensity of Use of Materials

Duan Ning*, Dan Zhigang

(National Cleaner Production Centre, Chinese Research Academy of Environmental Sciences, 100012, Beijing China)

Abstract: Scholars in many countries are again researching consumption of material and intensity of use (IU) of materials in relation to economic growth and technical capabilities. This renewed interest is the result of the standardization of indicators of economy-wide material flow analysis in developed countries. For the material consumption scale per capita researchers consistently consider it will initially increase and then decrease with the increasing income per capita, forming an "inverted U shape" Alternately, it will initially increase, and then stabilize at a relatively high value. For the IU it is expected to decrease continuously as income per capita increases. The "inverted U" environmental Kuznets curve (EKC) deeply impacted research into both material consumption and IU both before and after the indicators of economy-wide material flow analysis were defined. We analyzed data for 21 countries using linear, quadratic, and cubic polynomial models. Results show that as the economy develops, material consumption scale per capita increases; however IU may decrease or increase. The endogenous drive on material consumption scale per capita and EKC are totally different both in theory and practice. These results indicate the solution to sustainable development is the construction of a new material basis consisting entirely of recycled material for the economy and the development of technology in which solar energy is the primary driving energy.

Key words: Material consumption scale per capita; Direct material input; Intensity of use of materials; Circular economy

1. Introduction

Many scientists are again examining the relationship of material consumption scale and intensity of use (IU) of materials with economic growth after the EU proposed the standardized economy-wide material flow analysis indicators. Scholars in many countries have researched the development trends in material consumption scale per capita and IU, which is defined as the amount

---
* Corresponding author: Tel.: +86 10 8491 3945; Fax.: +86 10 8491 4626.
E-mail address: ningduan@craes.org.cn.

of material used per economic output, often measured as gross domestic product (GDP), as measured by the indicators of direct material input (DMI) and total material requirement (TMR). Researchers believe that trends in material consumption scale per capita will show inverted-U or inverted-L shapes with increasing income per capita. IU decreases with the increasing income per capita. However, results from our analysis on DMI and IU data from 21 countries showed that material consumption scale per capita increases as the economy develops. In addition, intensity of use of materials will decrease or increase with economic development. We demonstrate here that material consumption scale per capita follows a different trace against the inverted-U environmental Kuznets curve which has been universally followed and referred to by many. Finally we suggest that the solutions to resource depletion are to develop a circular economy whose material basis consists entirely of recycled material.

2. Background

We can discuss trends in material consumption scale and IU from different scholars in two stages. At the early stage, it was studied using indicators from the consumption of individual material such as cement, steel, and paper, or their combinations. In recent years, the indicators DMI and TMR have been applied to analyze trends in material consumption scale while the indicators of economy-wide material flow analysis were standardized in the European Union, USA, Japan, and other developed countries.

Three main opinions were generalized in earlier research on trends in material consumption. The first one considered that consumption of materials would decrease, as suggested by Malenbaum (1978), Ross (1981), Johansson (1983), Larson (1986), Williams (1987), Goldemberg (1988) and Jänicke (1997). The second option suggested that consumption of materials would increase, although it sometimes stayed stable or even decrease temporarily as proposed by Opschoor (1991), Hüttler (1999), Femia (2000), Jänicke (2001), de Bruyn (1997, 2002) and N. Duan (2004, 2005), who reanalyzed material consumption using data of a longer time period. The third option is transmaterialization. Labys (2002) pointed out that as the economy grew new and hi-tech materials would replace old ones, and consumption of material of any particular kind would go from small to large and then to small.

In this stage, intensity of use of materials was widely studied using as indicators consumption of a single bulk material and combinations of bulk materials. Malenbaum first described the hypothesis of the inversed U curve in IU. Williams (1987) provided as evidence for the hypothesis the consumption of cement, steel, and paper in the USA. Bernardini and Galli (1993) generalized Malenbaum's conclusions and suggested that IU for all kinds of materials follow a similar trend of first increasing, reaching a maximum point, then falling with economic growth. However, this conclusion could only explain trends in consumption of a single bulk material, and was not suitable for trends in consumption of mass bulk material, as pointed out by Considine (1991).

Recently, scholars turned to use DMI and TMR to study the relationship between material consumption and economic growth after the indicators for economy-wide material flow analysis were standardized in developed countries. Applying these indicators, they concluded that material requirements decreased or remained stable at high levels as economy grows. Canas et al. (2003), analyzing panel data for 16 countries econometrically concluded that as the economy grows, material consumption at first increased, but eventually declined after a certain high threshold arrived ,following a trend of inverted-U curve. Bringezu et al. (2004) suggested that a relative decoupling between GDP and DMI occurred in many countries, and followed an inverted-L curve.

There was also a consistent conclusion on IU as calculated by DMI and TMR. IU decreased with the growing economy, as pointed out by Tao (2003), who showed the declining trend in IU for the USA, Japan, Germany, and the Netherlands by DMI and TMR indicators from 1975 to 1994, and Šcasný et al. (2003) described a declining trend in IU for all countries in the European Union (EU-15) by DMI from 1980 to 2000.

3. Methods and Data

3.1. Methods

In this paper, country-specific analysis and overall analysis were used to discuss long-term tendencies in material consumption scale and IU in relation to economic growth. Country-specific analysis refers to analysis focusing on trends in a single country; overall analysis is defined as putting data from all countries together to find their general tendency. Importing and exporting materials among countries have become very universal because of the rapid development of an integrated global economy. The purpose of this paper is to study the overall relationship between material consumption scale and IU with the economic growth. Hence its conclusion will come from the overall analysis. However since the overall tendency of all the countries depends on the development tendency of the majority of the countries, or on the countries whose data have apparent impacts on the analysis, conclusions from country-specific analysis form an important reference for the overall analysis.

Linear, quadratic polynomial and cubic polynomial models are applied to perform the country-specific analysis and overall analysis using indicators DMI and IU. The regression equations used are shown as follows:

$$DMI = a + b_1 GDP + e \tag{1}$$

$$DMI = a + b_1 GDP + b_2 GDP^2 + e \tag{2}$$

$$DMI = a + b_1 GDP + b_2 GDP^2 + b_3 GDP^3 + e \tag{3}$$

Where DMI is DMI per capita, and GDP is GDP per capita, e is the error term; a, b_1, b_2 and b_3 are parameters to be estimated. When analyzing IU, the indicator DMI has to be substituted by the indicator IU.

The models assume the independent variable GDP per capita has covered effects generated by all the factors such as time, space, economy, policy and so on and hence is an integrated parameter. GDP represents impacts of technical level, changes in industrial structure, and regulatory actions on the indicators DMI and IU. The assumption is correct if the model passes a statistical test meaning using GDP as the only independent variable can explain impacts on DMI per capita and IU caused by changes in technical level, economic structure, policies and regulations and so on. If the model fails to pass the statistical test then GDP alone can not explain impacts on DMI per capita and IU caused by the factors.

3.2. Data

DMI and GDP data from China between 1990 and 2003 were calculated especially for a report on research into circular economy and the technologies of ecological industry (Duan et al., 2006). Data from all other countries were obtained from a figure describing the development of DMI/cap in the course of economic growth in the paper of Bringezu (2004), where the figure was digitized by WINDIG software to get the value of GDP and DMI and the error percentage is less than 0.5% as compared to the exact USA data for the period 1975-1994 (Adriaanse et al, 1997). IU data were calculated by the equation as follows: IU=DMI/GDP. GDP data for all countries were converted to constant prices and exchange rates for 1990. The data from the Czech Republic and Germany were eliminated because they are abnormal owing to a sudden shift in the regime in the Czech Republic and the reunion in Germany, as also pointed out by Bringezu.

In country-specific analysis, we only selected countries with data exceeding eight years including 16 countries such as USA, Japan, UK, etc., for the sake of comparing changing trend in each country in a single figure. We included 306 data points for DMI, GDP, and IU, respectively, from 21 industrialized and developing countries in the overall analysis.

4 Trend in material consumption scale per capita

4.1. Country-specific trends in DMI

4.1.1. Trends in DMI/cap for China

Fig. 1 shows the development of DMI/cap with the increasing GDP/cap in China. Results from linear, quadratic, and cubic models show that DMI/cap increases with increasing GDP/cap. As shown in Table 1, DMI/cap increases rapidly during the period 1990-2003; the DMI of China in 1990 is 2575 million tons, and 5452 million tons in 2003; representing a growth rate of 212%.

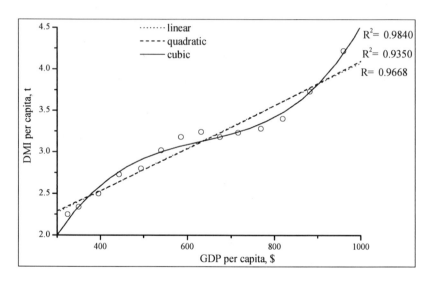

Fig. 1. Development of DMI/cap with the increasing GDP/cap in China

4.1.2. Country-specific trends in DMI /cap for some countries

Results of country-specific trends in DMI/cap using linear, quadratic, and cubic models are shown in Fig. 2, and Table 2 reports the statistical results of the models in Eqs. (1)-(3). From Table 2, it can be seen that the linear model for Netherlands and three models for Italy and the USA are of no statistical significance, and the values of the F-test for three models for all the other countries are superior to the critical value for the level of significance of 1%, 5%, or 10%, where we rejected the null hypothesis.

As shown in Fig. 2, there are 11 countries where the DMI per capita has a positive correlation to the GDP per capita. An exception is Ireland with negative correlation from results of the linear model. Australia, Belgium-Luxemburg, Finland, Denmark, Sweden, Portugal, Spain, Greece, and Norway have a 1% level of significance, and Japan and the UK show 5% significance. All of these demonstrate that DMI per capita of most countries increases with increasing GDP per capita.

Results from the quadratic model illustrated that the ascending trend of DMI per capita with increasing GDP per capita occurred in 10 countries: Ireland, Belgium-Luxemburg, Finland, Netherlands, Portugal, Japan, Sweden, Spain, Greece, and UK. These results can be considered a U shaped curve between DMI/cap and GDP/cap. For Australia, Denmark, and Norway, the trend is of an inverted-U curve resulting from the regression equation. However, the trend of DMI/cap in these countries is still on the upswing part of the inverted-U curve under the current GDP level as clearly shown in Fig. 2.

In the cubic model, a rising S shape is found between DMI per capita and GDP per capita in Portugal, Greece, Spain, Australia, Belgium-Luxemburg, Finland, Sweden, Denmark, and Norway, and a horizontal S shape accompanying a weak rise is seen in Ireland, Netherlands, UK, and Japan.

Table 1: DMI/cap of China from 1990 to 2003

Year	GDP/cap, $	DMI/cap, t/cap	DMI, Mtoe
1990	324	2.25	2575
1991	350	2.34	2713
1992	395	2.50	2925
1993	443	2.73	3231
1994	494	2.80	3351
1995	540	3.02	3652
1996	585	3.18	3888
1997	631	3.24	4008
1998	674	3.18	3970
1999	716	3.23	4065
2000	768	3.28	4156
2001	819	3.40	4345
2002	882	3.73	4796
2003	960	4.22	5452

As stated above, in the 16 countries including China there are 12, 14, 14 countries whose DMI per capita grew with increasing GDP per capita in the linear, quadratic, and cubic models, respectively. It can be pointed out that DMI per capita increase with rising GDP per capita in most countries.

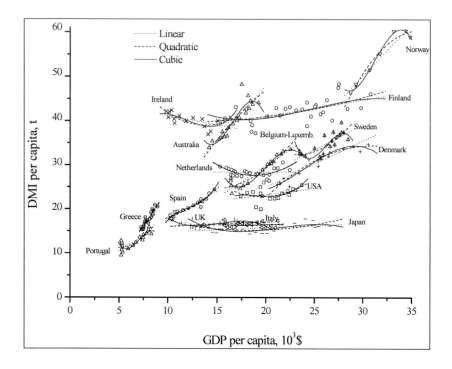

Fig. 2. Country-specific trends in DMI /cap

4.2. Overall trends in DMI per capita for all countries combined

Results of overall trends in DMI for all countries combined using linear, quadratic, and cubic models are shown in Fig. 3, and Table 3 presents the statistical results of the models in Eqs. (1)-(3). From Table 3, it can be seen that three models for all countries combined are of the level of significance of 1%. Results in Fig. 3 clearly indicate that all three models strongly support the conclusion that DMI per capita grows with increasing GDP per capita.

Comparing Fig. 3 with Fig. 1 and Fig. 2, it can be seen that the overall trend in DMI per capita for all countries combined is in good agreement with results from China and country-specific trends mentioned above.

Table 2. Estimation results of F-test for country-specific trends in DMI /cap

Country	Linear	Quadratic	Cubic
Ireland	5.9629**	6.5188***	4.7593***
Australia	61.7801***	29.4894***	23.5707***
Netherlands	0.0310-	5.8204***	3.6811**
Belgium-Luxemburg	140.7754***	88.6233***	102.9546***
Finland	14.5787***	7.3870***	4.9724***
Denmark	74.5000***	60.5990***	41.1642***
USA	2.0455-	1.7730-	1.2129-
Portugal	40.6041***	22.7750***	16.3277***
Japan	4.2697**	5.3062***	4.8866***
Sweden	24.0275***	13.4745***	12.1797***
Spain	203.3807***	111.7781***	87.8318***
Greece	102.5812***	48.1350***	30.0311***
Italy	0.0334-	0.1002-	0.0941-
UK	5.6296**	2.7336*	5.3481***
Norway	60.5706***	52.9281***	156.8564***

Dependent variable is DMI/cap (t). Independent variable is GDP/cap (US $ 1000-1990 prices and PPP). Regression parameters are not listed for length of the paper.

***Significant at the 1% level, ** Significant at the 5% level, * Significant at the 10% level, - No significance.

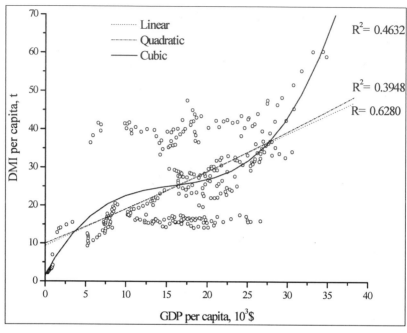

Fig. 3. Overall trends in DMI /cap for all countries combined

Table 3. Regression results of overall trends in DMI /cap for all countries combined

Model	Linear		Quadratic		Cubic	
a	9.3402	(1.2723)***	9.9480	(1.9014) ***	1.7882	(2.2240)
b1	0.9826	(0.0698) ***	0.8812	(0.2454) ***	3.8719	(0.5348) ***
b2	-	-	0.0032	(0.0075)	-0.2265	(0.0377) ***
b3	-	-	-	-	0.0048	(0.0008) ***
R^2	0.3944		0.3948		0.4632	
Adj. R^2	0.3924		0.3908		0.4579	
Number of samples	306		306		306	

Dependent variable is DMI/cap (t). Independent variable is GDP/cap (US $ 1000-1990 prices and PPP). *** Significant at the 1% level, Standard error of estimations inside round brackets.

5. Trend in intensity of use of materials

5.1. Country-specific trends in IU

5.1.1. Trends in IU for China

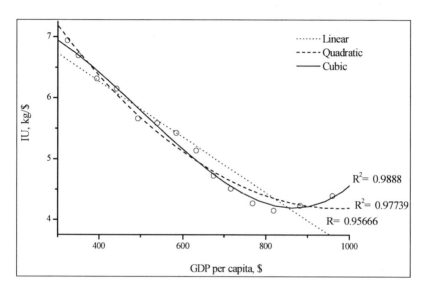

Fig. 4. Development of IU with the increasing GDP/cap in China

Fig. 4 presents the development of IU calculated by the DMI indicator with the increasing GDP/cap in China. Table 4 gives the IU of China from 1990 to 2003. Results from linear, quadratic, and cubic models show that IU decreases with increasing GDP/cap, as illustrated in Fig. 4. However, regression curves of both the quadratic model and cubic model turn upward at the end.

5.1.2. Country-specific trends in IU for some countries

Results of country-specific trends in IU using linear, quadratic, and cubic models are shown in Fig. 5, and Table 5 presents the statistical results of the models in Eqs. (1)-(3). Detailed results in Table 5 explain that the values of the F-test of three models for all countries are superior to the critical value for the level of significance of 1%, 5%, or 10%, being rejected for the null hypothesis, except that the linear model for Australia, Belgium-Luxemburg, Denmark, and Sweden are of no statistical significance. The quadratic model for Australia, Belgium-Luxemburg, Sweden, and the cubic model for Australia also have no statistical significance.

Table 4. IU of China from 1990 to 2003

Year	GDP/cap, $	IU, kg/$
1990	324	6.94
1991	350	6.70
1992	395	6.32
1993	443	6.15
1994	494	5.66
1995	540	5.59
1996	585	5.43
1997	631	5.14
1998	674	4.72
1999	716	4.51
2000	768	4.27
2001	819	4.16
2002	882	4.23
2003	960	4.40

As the linear regression curves shown in Fig. 5 demonstrate, the IU from nine countries shows negative correlation to the GDP per capita except for Greece and Norway. Fig. 5 also demonstrates that the current IU of most countries decreases with the increasing GDP per capita.

Results from quadratic model illustrated that IU from Ireland, Finland, Netherlands, USA, Portugal, Japan, Spain, Italy, and UK shows a U-shaped relation to the GDP per capita, and an inverted-U shaped relationship for Denmark, Greece, and Norway. Compared with results of the linear model, it is noted that those appearing as U-shaped relationships show an upward trend at the end of the curves, which may mean that IU would increase after its lower declining threshold.

In the cubic model, the trend in IU from Greece and Norway shows a continuous rise; development of IU from other countries declines with the increasing GDP per capita at a lower GDP/cap; after the GDP/cap reaches its medium value, the IU from Portugal, Spain, Belgium-Luxemburg, Sweden, and Demark remained stable temporarily, and other countries continued to fall. As for a higher GDP/cap, development of IU from Ireland, Spain, and USA shows an ascending curve, and others fall continuously.

Compared with results from three models, the linear model and cubic model give evidence that the IU of most countries continuously declines with the increasing GDP per capita. However, results of the quadratic model for most countries suggest that IU would increase after its lower drop limit.

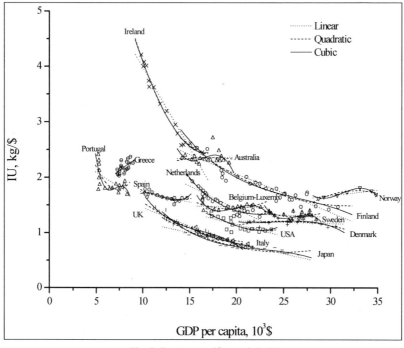

Fig. 5. Country-specific trends in IU

5.2. Overall trends in IU for all countries combined

Results of overall trends in IU for all countries using linear, quadratic, and cubic models are shown in Fig. 6, and Table 6 presents the statistical results of the models in Eqs. (1)-(3). From Table 6, it can be seen that three models for all countries are of the level of significance of 1%. Results in Fig. 6 clearly indicate that the linear model and cubic model show an overall trend in IU for most countries that is continuously declining with increasing GDP per capita. However, it is noted that results of the quadratic model for most countries suggest that IU declines at first until a certain threshold is reached, after which IU would increase with the rising GDP/cap.

Deep analysis on curves in Fig. 6 finds that the trend in IU is to decline with the increasing GDP per capita when GDP/cap is at a lower value, which is consistent in the three models. There are different developed trends for the three models after GDP/cap rises to a higher value. As for the quadratic model, when GDP/cap exceeds about $22,500, the quadratic curve rebounds upward, and the cubic model is relatively stable at GDP ranging from $15,000 to $30,000, but declines again at GDP/cap exceeding $30,000.

It is very interesting that the quadratic curve and cubic curve have an absolutely inverse changing trace when GDP/cap is higher than a certain threshold value. Such results can be found in the trend for China in Fig. 4 and Ireland, Finland, Netherlands, USA, Portugal, Japan, Sweden, Spain, Italy, and UK in Fig. 5.

As discussed above, we can see that the trend in IU is consistent no matter if the result is from analysis on China, country-specific, or overall consideration. Based on available data, it can be concluded that the overall trend in IU declines with increasing GDP per capita during the first stage. After the GDP/cap reaches a certain value, IU has two potential developing trends, that is, decrease or increase.

Table 5. Estimation results of F-test a for country-specific trends in IU

Country	Linear	Quadratic	Cubic
Ireland	217.3741***	340.9598***	209.0160***
Australia	0.0609-	0.0854-	1.0720-
Netherlands	159.6155***	197.5992***	126.6067***
Belgium-Luxemburg	0.2281-	2.1929-	5.6582***
Finland	174.8045***	135.3641***	95.6481***
Denmark	1.0477-	3.8537**	2.9235*
USA	16.6831***	9.9666***	6.3173***
Portugal	9.2323***	6.3258***	4.8470***
Japan	107.2550***	144.9621***	121.6502***
Sweden	0.7801-	1.4775-	2.6166*
Spain	23.5786***	14.4573***	11.4381***
Greece	10.6877***	5.0150***	3.1258**
Italy	90.6969***	47.5618***	30.2691***
UK	275.5209***	180.8718***	170.5482***
Norway	4.3209*	6.0406*	22.6711***

Dependent variable is DMI/cap (kg/$). Independent variable is GDP/cap (US $ 1000-1990 prices and PPP).

***Significant at the 1% level, ** Significant at the 5% level, * Significant at the 10% level, - No significance.

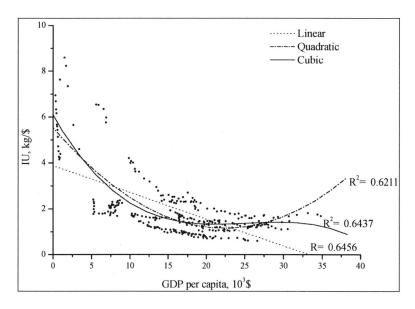

Fig. 6. Overall trends in IU for all countries

Table 6 Regression results of overall trends in IU for all countries combined

Model	Linear		Quadratic		Cubic	
a	3.8844 (0.1443)***		5.5338	(0.1738)***	6.0759	(0.2094)***
b1	-0.1167	(0.0079)***	-0.3919	(0.0225)***	-0.5906	(0.0503)***
b2	-	-	0.0088	(0.0007)***	0.0240	(0.0036)***
b3	-	-	-	-	-0.0003	(0.0001)***
R^2	0.4168		0.6211		0.6437	
Adj. R^2	0.4148		0.6186		0.6402	
Number of samples	306		306		306	

Dependent variable is IU (kg/$). Independent variable is GDP/cap (US $1000-1990 prices and PPP).

*** Significant at the 1% level, Standard error of estimations inside round brackets.

6. Results

6.1. About the general tendency in material consumption scale per capita

Results from our analysis show that the general tendency in material consumption scale per capita for both a single country and all countries combined is that it will increase gradually with economic growth. Occasionally, temporary stability follows the increase, which is absolutely different from the conclusions currently drawn by most scholars that material consumption scale per capita has already reached a relative constant level in most countries, and is even declining in some.

6.2. About the general tendency of IU

Another interesting finding involves the general tendency in IU based on DMI indicator which is to decline as a function of industrial structures, improvement on technology and other effective factors, as pointed out by all scholars. However, we suggest that overall trends in IU develop in two potential ways with the increasing GDP per capita, that is, a final decrease or increase based on available data.

6.3. About using GDP/cap as the only independent variable

Trends in DMI/cap and IU can be comprehensively explained by the independent variable GDP/cap, which was confirmed by overall analysis on all countries where the three models on both DMI/cap and IU have statistical significance at the level of 1%, as well as by results of country-specific analysis.

7. Discussion

7.1. Does material consumption scale per capita go along with the inverted-U shape curve?

The endogenous drive resulting in increment of material consumption scale per capita is the natural result of mankind's desire for a high standard of living. On the one hand, material consumption is an important element of a satisfied life in modern society. The desire for a high standard of living is rooted in the natural desire for growth. On the other hand, high environmental and ecological quality is also an important element of a satisfied life. Man is paying more and more attention to environmental and ecological quality with economic growth. Greater environmental and ecological quality requires a lower intensity of pollution discharge. It is well known that the more developed an economy becomes the higher its environmental criteria releases. As a result, the lower intensity pollution discharge is obtained, explaining why the EKC concentrates heavily on environmental quality due to human requirements. However, materials relating to contamination involved in EKC are the materials that people are unwilling to relinquish. According to their characteristics materials required to meet the desire for a high quality of living consist of two kinds of materials, one kind of material is such that mankind wishes it to increase with economic growth, such as all kinds of infrastructures. The other kind of material is such that mankind wishes it to decrease with the economic growth, such as all kinds of pollutions. So, the two kinds of materials reflect opposite driving forces in the course of economic growth, which has been neglected by almost all current research in this field. The two kinds of materials comply with different development traces because of the opposite driving forces. Such an endogenous drive results in material consumption scale per capita that rises gradually, and the pollution discharge per capita occurs in an inverted-U curve spontaneously.

At a lower GDP/cap level, people try their best to survive and say nothing of desire for environmental quality. When GDP/cap reaches a relatively higher level, it becomes easier for people to earn the materials for life and they express a stronger desire for environmental and ecological quality with economic growth. However, environmental improvement comes at a cost of mass input of materials. For example, waste water has always been discharged directly or received only primary treatment in the undeveloped economy. When the economy becomes developed, effluent disposal is undertaken by secondary treatment or tertiary treatment instead of primary treatment. Obviously, both cost in facilities and material input in maintenance from secondary or tertiary treatment are rather higher than that in the primary treatment. In other words, the decrease at the end of the EKC comes at the cost of mass material input. To realize the inverted-U EKC requires an accelerated increment of material consumption scale per capita. This point explains the opposite relationship between trends in material consumption scale per capita and EKC.

At present, significance of the so-called "step innovative technology" in material consumption scale per capita is neglected by many scholars in their research and they believe that material consumption scale per capita has already reached relative stability. Improvement of technology is the key approach to realizing a higher standard of living. Emergency of any step innovative technology will bring facilities and enjoyment to mankind. At the same time it provides mankind with feasible measures to meet their higher material demands. Historically, step innovative technology plays a significant role in determining economic growth and material consumption scale. Any time a new step technology appears it greatly improves the material life level and largely boosts the material consumption scale. For example, the steam engine in the 19th century drove the development of excavating and transporting facilities, as a result, consumption in coal resource increased.

7.2. Why do our suggestions differ from present conclusions?

Detailed comprehensive analysis was performed to study the development of the DMI/cap in the course of economic growth by Bringezu (2004) with linear, quadratic, and cubic models. Results of linear regression in his analysis showed that the slope is positive between the DMI/cap and the GDP/cap, supporting the conclusions in this paper. However, his paper focused on comparing country-specific data, which is different from our intention. It may be the one reason that Bringezu did not extensively analyze the overall trend in DMI/cap and the GDP/cap for all countries combined.

Analyzing the linear regression results from Bringezu (2004) shows that the DMI/cap of five countries: Norway, Australia, Sweden, Denmark, and Belgium-Luxembourg out of the seven highest DMI/cap exceeding the value of 30 t/cap are experiencing a strong positive relationship with GDP/cap. Additionally, the GDP of Norway, Finland, Sweden, Denmark, and Belgium-Luxembourg are all in excess of \$23,000. It is suggested that mass material consumption is still necessary for further improvement in economic growth in the countries with higher GDP/cap, and to a large extent, implies that economic development will rely intensively on material consumption in the future.

The inverted-U curve between material consumption per capita and economic growth was concluded by Canas (2003). But his analysis did not include data from Norway, Australia, and Belgium-Luxembourg whose DMI/cap shows a strong positive relationship to GDP/cap and both DMI/cap and GDP/cap are higher than others. So, it is worth discussing the conclusions from Canas.

7.3. What relationship exists between material consumption per capita and IU?

According to discussions in Sections 4 and 5, the overall trend in material consumption scale per capita and IU with the increasing GDP/cap can be illustrated by the models in Fig. 7. One model is that material consumption scale per capita increases with the quadratic or cubic curve continuously, and the trend in IU declines initially, but finally appears to have the same trend as material consumption scale per capita with the rising GDP/cap as shown in Fig. 7(a). Another model is that material consumption scale per capita increases with the cubic curve continuously, and IU develops as an opposite trend as shown in Fig.7 (b). If the trend in IU does not decline continuously, as suggested by most scholars, but rises at the end of curve as occurred in the first model, it might be too optimistic to predict material consumption scale per capita and its developing trace from current results, as a result, the time to depletion of existing mineral resources would come earlier than anticipated.

Material consumption scale per capita is a product of IU multiplied by GDP/cap. Though IU has a direct effect on material consumption scale, a drop in IU does not necessarily lead to a decline in material consumption scale per capita. This is explained by trends in DMI and IU from China, Ireland, Denmark, Finland, Netherlands, USA, Portugal, Japan, and UK that show decreasing IU followed by increasing DMI/cap with the increment of GDP/cap clearly resulting from comparing

Fig. 1 with Fig. 4, and Fig. 3 with Fig. 6. Models in Fig. 7 indicate that the final development trace of DMI/cap increases with rising GDP per capita whether the trend in IU declines or increases. It can be seen that the increment in DMI/cap is due to the rising GDP/cap, which is faster than that of the reduction from the falling IU. This causes the result mentioned above, that is, the reducing speed in IU can not catch up with the increasing speed in GDP/cap.

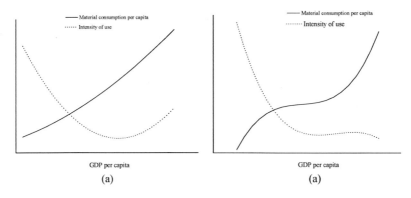

Fig. 7. Models for overall trend in material consumption per capita and IU with the increasing GDP/cap

7.4. What solutions solve the contradiction between the finite resource reserves and the infinite economic growth?

Conclusions in this paper firmly indicate that the overall trend in material consumption scale per capita can increase continuously, however, fossil resource reserves become more and more scarce. What solutions solve the contradiction between the finite resource reserves and the infinite economic growth? To realize sustainable development we must concentrate not only on material consumption scale, but also on evolutionary tendency in material basis for economic growth, and focus on restructuring our material metabolism to regulate material basis.

In fact, material requirements are driven by demand for function that uses material as carrier. Technically, many plants can be used to produce seed oil that can not only refine petrol to solve transportation fuel problems, but also can be used as raw materials for petrochemical engineering to fabricate fertilizers, plastics, chemicals, fibers and other products. Furthermore, high performance materials equal to iron, steel, and other metals can also be synthesized from seed oil with advanced petrochemical engineering technology. These examples show that primary raw material to meet life-style requirement can be supplied by renewable biomass resources based on advanced technology. That means that economic growth will depend on renewable biomass resources, namely recycling concepts in a circular economy. Recycling concepts also include recycling of materials involved in downstream processes using wastes of upstream processes as raw materials, and recycling industrial substances such as steel scrap, aluminum scrap, and discarded building materials, as well as cascading use of energy and water resources by both citizens and industries. A prerequisite for recycling material is a sufficient supply of energy. A circular economy predicts that all primary driving energy for society would come from solar energy and advanced technology would be developed to use solar energy efficiently.

Overall plans for solving the issue of depletion of earth resources by circular economy focus on increasing efficiency of material use, maximum recycling of materials, increasing the proportion of renewable materials consumed, reducing and finally ceasing use of fossil resources. As a result, a new material basis of economy based on entirely recycled material is constructed where a batch of material is recycled many times, which means that many batches of materials are used to meet the requirements of people. This would be a feasible approach to solving the contradiction between finite resource reserves and infinite economic growth.

8. Conclusions

(1) The time when existing mineral resources become depleted could come earlier than anticipated due to the continuously increasing material consumption scale with economic development. Mankind should consider this deeply, both mentally and technically.

(2) Given the possible trend in resource exhaustion, the solution to this issue is that a new material basis of economy based on entirely recycled material is constructed and advanced technology is developed in which solar energy is the primary driving energy; resulting in a circular economy.

(3) Further studies with more extensive data should be conducted to analyze if trends in IU with economic growth follows U-shaped quadratic curves or declining cubic curves. Comparisons and intersection among DMI/cap, IU, and inverted-U EKC will be studied to find their interconnection and obtain their quantified relationships from which new approaches to solving environmental problems will be developed based on material flow analysis.

Acknowledgements

This work was supported by the National Key Project 2003BA614A-02, Ministry of Science and Technology, China. We thank anonymous referees for their helpful comments and technical support on this paper.

References

[1]   Adriaanes, A., Bringezu, S., Hammond, A., et al., 1997. Resource flows: the material basis of industrial economics. World Resources Institute, Washington, DC.

[2]   Bernardini, O., Galli, R., 1993. Dematerialization: long-term trends in the intensity of use of materials and energy. Futures 25, 431-447.

[3]   Bringezu, S., Schütz, H., Steger, S., 2004. International comparison of resource use and its relation to economic growth: the development of total material requirement, direct material inputs and hidden flows and the structure of TMR. Ecological Economics 51, 97-124.

[4]   Canas, A., Ferrão, P., Conceicão, P., 2003. A new environmental Kuznets curve? Relationship between direct material input and income per capita: evidence from industrial countries. Ecological Economics 46, 217-229.

[5]   Considine, T.J., 1991. Economic and technological determinants of the material intensity of use. Land Economics 67(1), 99-115.

[6]   De Bruyn, S.M., 2002. Dematerialization and rematerialization as two recurring phenomena of industrial ecology. In: Ayers, R.U., Ayres, L. (Eds.), Handbook of Industrial Ecology. Edward Elgar Publishers, Cheltenham, pp. 209-222.

[7]   De Bruyn, S.M., Opschoor, J.B., 1997. Developments in the throughput–income relationship: theoretical and empirical observations. Ecological Economics 20, 255-268.

[8]   Duan N., Qiao Q., Sun Q.H., et al., 2006. A report on researches in circular economy and technologies of ecological industry. Chinese Research Academy of Environment Sciences, Beijing. (in Chinese)

[9]   Duan N., 2005. Material metabolism and circular economy. China Environmental Science 25(3), 320-323. (in Chinese)

[10] Duan N., Deng H., 2004. "Upward-multipeaks theory" and circular economy. World Nonferrous Metals (10), 7-9. (in Chinese)

[11] Duan N., Deng H., 2004. "Upward-multipeaks theory" and circular economy (Continuous). World Nonferrous Metals (11), 9-12. (in Chinese)

[12] Femia, A., 2000. A material flow account for Italy 1988. Eurostat Working Papers 2/2000/B/8, Luxembourg.

[13] Goldemberg, J., Johansson, T.B., Teddy, A.K.N., et al., 1988. Energy for a sustainable world. Wiley Eastern.

[14] Hüttler, W., Schandl, H., 1999. Are industrial economies on the path of dematerialization? Material flow accounts for Austria 1960-1996: indicators and international comparison. In: Third ConAccount Meeting: Ecologizing Societal Metabolism, Amsterdam.

[15] Jänicke, M., 2001. Towards an end to the "Era of Materials"? Discussion of a hypothesis. In: Boinder, M., Jänicke, M., Petschow, U. (Eds), Green Industrial Restructuring, pp. 45-57

[16] Jänicke, M., Moch, H., Boinder, M., 1997. "Dirty industries": patterns of change in industrial countries. Environmental and Resource Economics 9, 467-491.

[17] Johansson, T.B., Steen, P., Bogren, E., et al., 1983. Sweden beyond oil-the efficient use of energy. Science 219, 355-361.

[18] Labys W.C., 2002. Transmaterialization. In: Ayers, R.U., Ayres, L. (Eds.), Handbook of Industrial Ecology. Edward Elgar Publishers, Cheltenham, pp. 202-208.

[19] Larson, E.D., Ross, R.H., Williams, R.H., 1986. Beyond the era of materials. Scientific America 254(6), 24-31.

[20] Malenbaum. W., 1978. World demand for raw materials in 1985 and 2000. McGraw-Hill, NewYork.

[21] Opschoor, J.B., Reijnders L., 1991. Towards sustainable development indicators. In: Kuik. Verbruggen.(Eds), In Search of Indicators of Sustainable Development. Kluwer Academic Publishers, Boston, pp. 7-28.

[22] Ross, M., Purcell, A.H., 1981. Decline of materials intensiveness-the US pulp and paper industry. Resources Policy 7, 235-250.

[23] Šcasný, M., Kovanda, J., Hak, T., 2003. Material flow accounts, balances and derived indicators for the Czech Republic during the 1990s: results and recommendations for methodological improvements. Ecological Economics 45, 41-57.

[24] Tao Z.P., 2003. Eco-rucksack and eco-footprint-weight and area conception of sustainable development. Economic Science Press, Beijing. (*in Chinese*)

[25] Williams, R.H., Larson, E.D., Ross, R.H., 1987. Materials affluence and industrial energy use. Annual Review of Energy 1987 12, 99-144.

Renewable Energy Strategy for Eco-industrial Development in China

Li Xingguang[1]

(1. School of Management, Dalian University of Technology, 2.PetroChina Dalian Marine Shipping Company ; Liaoning, 116001)

Abstract: Industries are playing a more and more important role for Chinese sustainable development. Eco-industrial parks (EIP) development is one of the most prosperous strategies to achieve sustainable industrial development. As one of the key aspects for EIP, energy consumption and renewable energy (RE) development strategies are mainly discussed in this paper. Suggestions are given in terms of how to promote RE in EIPs in China in a long-term run.

Keywords: Renewable Energy, Eco-industrial parks, Sustainable Development

1 Introduction

China is the most populous country in the world and the second largest energy consumer after the United States. Production and consumption of coal, its dominant fuel, is the highest in the world. Coal makes up 65% of China's primary energy consumption, and China's coal consumption in 2003 was 1.53 billion short tons, or 28% of the world total.

One of the key strategies in China's Circular Economy (CE) initiative is to use eco-industrial parks to help generating a much higher productivity and efficiency of resource utilization. The Circular Economy approach to resource-use efficiency integrates cleaner production and industrial ecology in a broader system encompassing industrial firms, networks or chains of firms, eco-industrial parks, and regional infrastructure to support resource optimization [1]. There are two kinds of solutions that are suitable for China to build an eco-industrial park. One is to redesign the existent industrial parks and build the network of byproduct exchange to connect the enterprises with each other so that the park can achieve the low energy and material consumption. The other is to build a new eco-industrial park with one or several anchor companies.

There are some successful EIPs in China in the past few years. For instance, Zaozhuang EIP initiative in Shandong Province in North China, with key feature of transforming a traditional

---

[1]   Li Xingguang is an accountant for PetroChina Dalian Marine Shipping Company and also a Ph.D candidate in School of Management, Dalian University of Technology. Email: uussaa911@hotmail.com.

industrial zone to eco-industrial park; and Nanhai EIP initiative in Guangdong Province in South China, with key feature of developing environmental protection industry in a green field.

2 Renewable Energy Developments: Advantages and Difficulties

Renewable energy can not totally replace the conventional fuels in the industrial park due to the technology. Renewable energy can only be used as an assistant energy resource in the industrial park, but it can reduce the fossil fuel consumption and environmental impact. There are also some successful examples in the world, for instance the Dyfi Valley Community Renewable Energy Project which is located in United Kingdom. This project aims to benefit the community's 12,500 or so residents by encouraging the local people to engage with energy issue, establishing some community-based renewable energy installation. Furthermore, the project wanted to improve understanding and support for renewable energy by maximizing the local benefits. The total installed capacity of completed renewable energy schemes was 205 kW electrical capacity (hydro, wind, solar) and 150 kW heat capacity (solar, wood, heat pump). And now the people in the valley do not need to buy any energy from the other companies outside the park.

As we analyzed, there are two types of Industrial Parks that are suitable for China Comparing these two types of IP, the second one is easier to achieve because the most important and difficult problem is to build deep trust among the companies and the investment for the construction of exchange network is also a problem for the first type. Normally the investment will be provided by the local government. But for the second type, the anchor company can provide the technology and finance support for the eco-industrial park. At the same time, the company can also apply for the finance support from the government because this industrial park is a good sustainable solution for China's circular economic. Furthermore, Himin Group is a great company whose main production is the solar energy system in China, which is another advantage to build the IP with renewable energy as one of the energy resource.

For advantages, these two kinds of industrial parks can both reduce fossil fuel consumptions. In the eco-industrial park, energy can be saved by increasing energy use efficiency, but in the industrial park with solar energy, the fossil fuel consumption can be reduced by replacing solar energy. Furthermore, they are both friendly to the environment. For disadvantages, the eco-industrial park is much harder to achieve than the industrial park with solar energy.

Eco-industrial parks have been primarily described in the industrial ecology literature as a means of managing material and energy flows with attention to the possibility of particular chemical linkages. So the companies located in the park must have the capability to be connected by the industrial network according to the chemical linkages. These companies are sometimes hard to find and to be collected together to build an eco-industrial park. Some companies are not used to work "in community" and may fear the interdependence this creates.

On the other hand, the old Industrial Parks IPs have been aroused to make rational adjustment in the distribution of small and medium enterprises (SMEs) in towns or cities in China since 1980s. Over the years two kinds of IPs have been developed in the country. The first are parks with one or several anchor companies. The second are parks composed of SMEs from many distinct business sectors but without anchor enterprises, which account for the main part of the total IPs.

The challenges of guiding the second IPs onto eco-industrial parks might be that a series of obstacles, e.g. irregular layout of enterprises site, less available mass-byproduct-water for reuse, short of the conditions for energy cascading use and weak in information exchange among the enterprises and park administration. Furthermore, companies using each other's residual products as inputs face the risk of losing a critical supply or market if a plant closes down. This will break the whole industrial network so that the park will not survive any more, which will lead to a big economic loss.

3 Conclusions and Strategies

From the above discussion, China's energy consumption development follows the developed countries' experience. The change of energy supply structure, however, is not exactly the same. The structure of energy supply in developed countries generally experienced two transformations: from firewood to coal, and from coal to petroleum. The developed countries are now moving toward diversification of energy for the third transformation, where the renewable energy resources will prosperously substitute the fossil energy to become the dominant resource [2]. For the time being, China has completed its first energy transformation, just entered the rapid development of oil and natural gas. Energy diversification structure is far from being formed. The energy situation that China is facing, however, is very different from the developed countries. Considering the conditions of the domestic energy resources, the unstable international oil market, as well as the deterioration of the environmental pollution problem, China has difficulties to complete the second energy revolution in the similar way that the developed countries experienced, to form an energy consumption structure, which is dominated by oil and natural gas. In other words, China will remain a coal-dominated energy structure for quite a long time, which will make the air pollution even worse. Vicious circle shaped as demands pressures result in more resource pressures, and resource pressures result in more environmental pressures, owning to lower energy efficiency indeed. Therefore, China must jump over the second energy revolution, and should focus on the development of renewable energy, which is the strategic choice for China to sustain energy supply and environmental carrying.

Reference:

[1] Ernest A Lowe, (2001), Eco-industrial Park Handbook for Asian Developing Countries, Report to Asia Development Bank, Indigodv.Com, URL: http://www.indigodev.com/ADBHBdownloads.html (2005-9-9

[2] Project team of macro-economy research in National Ministry of Planning, "Mid and Long-term Energy Strategy for China," Beijing: Chinese Planning Publication, Feb. 1999.

A Comparison of Recycling Management Models of Waste Electrical and Electronic Equipment

Lan Ying

(School of Management, Dalian University of Technology, Dalian , P.R.China, 116024)

Abstract: Waste electrical and electronic equipment (WEEE) has been an emerging problem in recent years. The potential environmental hazards of WEEE and the loss of renewable resources highlight the demand for the increased recycling of WEEE on a world-wide scale. Recycling of WEEE can not only reduce environmental pollution, but also conserve resources. This paper compares several recycling management models of WEEE in developed and developing countries, and analyzes the problems of these models. In the end, several suggestions are provided for the establishment of WEEE recycling management systems in China.

Key words: Waste electrical and electronic equipment (WEEE); Management model; Recycling

1 Introduction

Waste electrical and electronic equipment (WEEE) has been an emerging problem in recent years. WEEE contains lots of toxic materials, such as lead, mercury, cadmium, hexavalent chromium, and brominated flame-retardants, which result in significant environment pollution and do harm to human health. On the other hand, there are many reusable materials and precious metals such as gold, Cu and plastic in WEEE. The potential environmental hazards resulting from the disposal of WEEE, such as the accumulation or leaching of hazardous and toxic substances, and the loss of renewable resources such as scarce and precious metals highlight the demand for the increased recycling of WEEE on a world-wide scale.

To address the environment problem of WEEE, developed countries enacted several laws to manage the recycling of WEEE, carried out relevant practices, and acquired good effects. Relevant research on the recycling management of WEEE was also done, for example the system design of WEEE recycling management [1], recycling management models [2], influencing factors of WEEE recycling management [3] and recycling technologies [4]. Lots of valuable results were obtained. Many studies toward WEEE recycling management were made in China. However, most of them concentrated on the summary of good practices in developed countries [5-6]. In order to introduce WEEE recycling management more intensively a comparison of several recycling management models of WEEE is drawn. This paper is organized as follows. The next section introduces the concept and characteristics of WEEE. Then, the recycling management models of WEEE in typical developed countries are compared in detail in section 3. Section 4 presents some suggestions about the establishment of recycling management models of WEEE in China. In the end, a conclusion is given.

2 Definitions and characteristics of WEEE

2.1 Definitions of WEEE

WEEE, which is also named as 'electronic waste' or 'e-waste' for short, is a generic term embracing various forms of electric and electronic equipment(EEE) that have no useful value to their owners. Actually, there is no uniform definition up to the present. OECD's definition is as follows: any appliance using electric power that has reached its end-of-life.

WEEE is classified into ten categories according to the EU Directive: (1) large household appliances; (2) small household appliances; (3) IT and telecommunications equipment; (4) consumer equipment; (5) lighting equipment; (6) electrical and electronic tools (with the exception of large-scale stationary industrial tools); (7) toys, leisure and sports equipment; (8) medical devices (with the exception of all implanted and infected products); (9) monitoring and control instruments; (10) automatic dispensers.

2.2 Characteristics of WEEE

Lots of WEEE is produced by developed countries. The United States of America are the biggest generator/producer of WEEE. About 6 million tons of WEEE is generated per year. WEEE already accounts for 2-5% of the municipal solid waste (MSW) and is still growing rapidly [7]. In Europe, about 6 million tons of EEE is discarded each year [8]. About 600,000 tons of household appliances accounting for 1% of MSW are discarded in Japan resulting in serious environment pollution [9].

A report estimated that business and individual households make approximately 1.38 million personal computers obsolete every year in India [10]. At present, a great quantity of waste household equipment is retained in China. Most of these appliances came into households 20 years ago and millions of them are to be discarded every year, such as 5,000,000 TV sets, 5,000,000 washing machines and 4,000,000 refrigerators [11].

WEEE has significant negative impact on the environment and human health when it is disposed of without any control. These discarded products contain many toxic materials such as lead, mercury, cadmium, hexavalent chromium and halogenated flame retardants. Burning them will create dioxins emissions. About 70% of the heavy metals in municipal solid waste come from

WEEE. These toxins can cause brain damage, allergic reactions and cancer [12]. Therefore, WEEE belongs to hazardous solid waste and its danger is potential and chronic.

At the same time, WEEE contains a certain quantity of valuable materials such as precious metals. The report in [13] showed that a ton of electronic board cards collected at random contain 271.8 kg plastic, 129.6 kg copper, 0.5 kg gold, 40.8 kg ferrous, 19.9 kg tin and so on. Therefore, recycling of WEEE is becoming an attractive business.

3 Recycling management models of WEEE

Some relevant practices of WEEE recycling management have been carried out in developed countries with good effects. WEEE recycling management models can be divided into two types: the mature recycling management models toward the collection, recycling and disposal of WEEE in developed countries and the exploratory recycling management models in developing countries in order to minimize the negative environment impacts of WEEE.

3.1 Management models in developed countries

Developed countries enacted several laws to manage the recycling of WEEE, such as WEEE Directive and RoHS Directive. The European Union's WEEE Directive makes it incumbent on manufacturers and importers in EU states to take back their products from consumers and ensure environmentally sound disposal. It has one of the most comprehensive WEEE schemes exists in the Netherlands. The Dutch government approved the "disposal of white and brown goods decree" in 1997, and expanded it twice in 1999 to include virtually all electrical and electronic household appliances, with the exception of lighting. Some elements of the scheme include: retailers and local authorities are responsible for collecting (or financing collecting) of WEEE, but do not need to incur any costs for further disposal and recycling; producers/importers have the responsibility to accept the collected apparatus without charge, and to transport and recycle WEEE; only non-recyclable material can be landfilled[14]. In a word, most recycling management models of WEEE in Europe are established according to the extended producer responsibility (EPR) principle. The cost relative to recycling is paid by manufacturers. Consumers just need to bring obsolete appliances to the designated collection points and do not have to pay extra recycling fees.

In April 2001, the Electric Household Appliance Recycling Law (EHARL) took effect in Japan and the responsibilities of recycling four electric household appliances (refrigerators, washing machines, air conditioners and TV sets) are shifted from local governments to producers. Retailers are responsible for transporting WEEE to producers' stock-points after receiving them from final users. Consumers need to carry discarded appliances to the designated collection points or contact the designated recycling companies. Consumers should pay relevant recycling fees when they return WEEE to producers (mainly through retailers) [15]. Material recycling targets set by EHARL are 50% for refrigerators and washing machines, 55% for TV sets and 60% for air conditioners. Because of the explicit responsibilities of stakeholders, the recycling of WEEE has obviously effects.

Motivated by the adoption of the EPR principle, the United States of America began to manage the recycling and reusing of electronic products. The National Electronics Product Stewardship Initiative (NEPSI), proposed by the Environmental Protection Agency (EPA) in 1999, tried to set up a national recycling system with a focus on TVs and personal computers. Some states are moving ahead with the programs and regulations of their own. For example, the Minnesota Office of Environmental Assistance (OEA) carried out an electronics recycling demonstration project over 1999 and 2000 in partnership with Sony Electronics Inc., Waste Management- Asset Recovery Group, Panasonic- Matsushita and American Plastics Council. Other states, such as Massachusetts, California and New York, have either put legislation in place to ban cathode ray tubes from landfill or are planning [16].

Generally speaking, the major differences of the recycling models of WEEE in developed countries are the responsibilities of stakeholders, such as manufacturers, retailers and consumers. Table 1 shows the differences of three stakeholders' duties in EU, Japan, and USA and presents the advantages and disadvantages of the WEEE recycling models.

Table 1　A Comparison of Electronic Waste Recycling Models in Typical Developed Countries

Model	Responsibilities of manufacturers	Responsibilities of retailers	Responsibilities of consumers	Advantages	Disadvantages
Model of EU	Dispose of WEEE and pay the take-back cost	Take back products and deliver them to producers	Take WEEE to the designated collection points	Producers pay the cost; higher collection rate	Producers transfer the recycling cost to consumers
Model of Japan	Dispose of WEEE	Take back products and deliver them to the designated recycling companies	Pay the relevant cost of transportation and disposal	Consumers pay the cost and producers participate actively	Rely on the participation consciousness and behavior of consumers and may result in discard illegally
Model of USA	Participate voluntarily	Participate voluntarily	Participate voluntarily	Local governments and stakeholders share the cost of collection and disposal	Low recycling rate and a lot of WEEE is exported

3.2 Management models in developing countries

Although the quantity of indigenous WEEE in every region is still relatively small, populous countries such as China and India have undoubtedly become huge producers of e-waste in absolute terms. Combined with the existence of a low-income informal sector, many developing countries permit a profitable WEEE recycling business thriving on uncontrolled and risky low-cost techniques. Most of the participants in this sector are not aware of environmental and health risks or do not know better practices. In China and India, one kind of complex WEEE recycling infrastructure based on and executed by a very informal sector has developed, which reflects a long tradition in waste recycling. Rag pickers and waste dealers are easy to adapt to the waste stream of WEEE and a large number of new businesses are created in the process of reusing components or extracting secondary raw materials. Besides, business-driven WEEE recycling systems come about without any government intervention.

China is a resource-poor country, with per capita distribution of natural resources at 58% of the world average [17]. In addition, large amounts of raw materials and components are needed because of the rapid economic progress. Thus, recycled materials have more market spaces in China than in Europe or North America. The WEEE recycling industry also provides the opportunities to increase income for both individuals and enterprises. Low risk processing, such as the manual dismantling of WEEE, can offer some job opportunities for low and medium skilled laborers if given proper training. However, some recycling processing is extremely harmful and needs to transfer to formal industries. Moreover, these developing countries including China are facing other difficulties like illegal imports.

4 Suggestions for a Chinese model

In order to design an appropriate management model of WEEE, Chinese governments and enterprises should learn from the recycling experiences of developed countries. At the same time, the recycling management system of WEEE should be designed according to the present situation in China, not just simply copy other countries' models. In addition, because of the different cultural background, it will bring serious results if these models are adopted directly in China. The current WEEE recycling system is composed of many informal organizations and has existed for several years. Therefore, several suggestions are presented according to the experiences of developed countries and the actual situation in China.

(1) Strengthen legislation on WEEE management

It is important to regulate the responsibilities of manufacturers by legislation. There is no relevant legislation about the EPR principle in China. Most EEE producers do not have the responsibilities to recycle or dispose of WEEE. However, under the shock of the WEEE Directive, some pilots begin to pay attention to the recycling of WEEE. Therefore, it is time to formulate pertinent regulations to strengthen the EPR principle and stipulate producers to answer for the whole lifecycle of EEE. It also urges companies to improve product design and explore eco-friendly products in order to control waste generation from source.

(2) Establish regular WEEE recycling systems

The current recycling system of WEEE is characterized by its individual collection model. Though this model is helpful to raise the collection rate of WEEE, most people may refuse individual organizations to enter their homes because of security, which is not helpful to collect WEEE, especially big WEEE. Therefore, it is a right solution to incorporate the informal sector into a well-regulated WEEE management system and to issue collection certificates. In addition, retailers also can take back WEEE by offering old-for-new service.

(3) Reinforce macro-management of government

Governments should guide the recycling management of WEEE by relevant legislations and regulations. The concrete methods are: (a) to present preferential revenue policies to regular recycling companies; (b) to strengthen environment supervision of WEEE; (c) to arouse people's participation by publicizing the hazards of WEEE.

5 Conclusions

Environment and resource problems make every country attach importance to the recycling management of WEEE, which further hastens governments to develop more reasonable WEEE management models. Moreover, WEEE recycling can bring environmental, economical, and social profits to promote sustainable development. In China, some trial schemes toward WEEE recycling management have been carried out, but that is not enough to solve the problems of WEEE recycling management at present. Therefore, it becomes a crucial problem how to incorporate the informal sector into a regular WEEE management system and to develop an appropriate management model in China on the basis of learning the advantages of recycling management models of developed countries.

References

[1] Sinha-Khetriwal D, Kraeuchi P, Schwaninger M. A comparison of electronic waste recycling in Switzerland and in India[J]. Environmental Impact Assessment Review, 2005, 25(5):492–504

[2] Gottberg A, et al. Producer responsibility, waste minimisation and the WEEE Directive Case studies in eco-design from the European lighting sector[J]. Science of the total environment, 2006, 359(1-3):38-56

[3] Streicher-Porte M, Widmer R, et al. Key drivers of the e-waste recycling system: Assessing and modelling e-waste processing in the informal sector in Delhi[J]. Environmental Impact Assessment Review, 2005, 25(5):472–491

[4] Cui J, Forssberg E. Mechanical recycling of waste electric and electronic equipment: a review[J]. Journal of Hazardous Materials, 2003, 99(3):243-263

[5] Chen Su, Fu Juan, Chen Chaomeng. A Study on Current Treatment and Management of Electronic Waste[J]. Journal of Nanhua University(Science & Engineering Edition, 2003,17(1):81-85 (in Chinese)

[6] Liu Mufan, et al. Responsibility Extending System of Producer and Control Measure in Electronic Waste Management[J]. Science & Technology Progress and Policy, 2005, (2):57-59. (in Chinese)

[7] Wang Jingwei, Shi Dehan, Chen Xulian. The Status Quo of Waste Electric and Electronic Equipment Recycling Industry in USA[J]. Shanghai Environmental Science, 2003, 22(12): 1034-1037 (in Chinese)

[8] Cooper T. WEEE, WEEE, WEEE, WEEE, all the way home? An evaluation of proposed electrical and electronic waste legislation[J]. European Environment, 2000, 10(3):121-130

[9] Onorato D. Japanese Recycling Law Takes Effect[J]. Waste Age, 2001, 32(6):25-26

[10] Agarwal R, Ranjan R, Sarkar P. Scrapping the hi-tech myth: computer waste in India. New Delhi: Toxic Link; 2003

[11] Mao Yuru, Li Xing. Preliminary exploring current condition and recycling system of electronic waste[J]. Recycling Research, 2004, (2):11-14 (in Chinese)

[12] Puckett J, Smith T. Exporting harm: the high-tech trashing of Asia The Basel Action Network. Seattle: Silicon Valley Toxics Coalition; 2002

[13] Seaver W.B. Design Considerations for Remanufacturability, Recyclability and Reusability of User Interface Modules. IEEE, 1994(5):241-245

[14] Mayers K, France C. Meeting the 'Producer Responsibility' Challenge: The Management of Waste Electrical and Electronic Equipment in the UK, GMI 25, Spring 1999:51-66

[15] Terazono A, Yoshida A, et al. Material cycles in Asia: especially the recycling loop between Japan and China[J]. Journal Material Cycles waste management, 2004, 6(2):82-96

[16] Jofre S, Morioka T. Waste management of electric and electronic equipment: comparative analysis of end-of-life strategies[J]. Journal Material Cycles Waste Management, 2005, 7(1):24-32

[17] Hicks C, Dietmar R, Eugster M. The recycling and disposal of electrical and electronic waste in china legislative and market responses[J]. Environmental Impact Assessment Review, 2005, 25(5):459–471.

Studies on Extended Producer Responsibility Scenarios Based on Analysis of Negative Values of End-of-Life (EOL) Products

Zhao Yiping

(School of Management, Dalian University of Technology, 116023, Liaoning, P.R.C.)

Abstract: Economic profitability and negative ecological influence of end-of=life (EOL) products are reversely correlated. In this paper, economic and ecological values contained in EOL products are decomposed according to the real recycling process at the post-consumption stage. Four crucial scenarios while implementing EPR are classified and studied based on which principal rules for implementing EPR and key influential factors for EPR design are identified and analyzed. Current situations and potential barriers for China are studied accordingly and policy implications are discussed at last.

Keywords: EOL products; Extended Producer Responsibility; Scenarios; Negative Values

1 Introduction

With the rapid development of productions and consumptions in modern economies, end-of-life (EOL) products are causing more and more serious environmental pressure. Parts of the EOL products are collected, recycled and disposed by the market but usually with an unsatisfactory recycling rate, for their components were not designed to be easily separated and the adopted materials cannot be easily recycled. On the contrary, large amounts of those EOL products are reluctant to be accepted or refused by the market for their poor recycling profitability. These EOL products such as used cars and household equipments are discarded and dumped into the nature by their last owners, not only occupying quite a lot of land space, but also leading to earth and water pollution and wasting massive useful natural resources.

Facing the growing pressures caused by EOL products especially the dumped EOL products some waste management practitioners suggested a new solving principle, namely extended producer responsibility (EPR). After being formally forwarded in 1990[1], PR has been widely accepted and applied in over 20 countries and regions in the world. It proves that EPR can effectively internalized the external costs of EOL products to the beginning of the production system and provide incentives to the producers to improve the products' environmental performances throughout their whole life cycle by eco-designing and other source management measures[2].

For a successful EPR program reasonable allocation of responsibilities especially the economic ones is one of the key elements and also the most important challenge for policy makers [3]. Even though a sound allocation of economic responsibility is so important to the success of EPR programs, there are few studies done on the policy guidance or the underlying principles. In practice, each country applying EPR is taking different solutions. In Japan part of the separating costs are assigned to the producers and fixed recycling fees are set for the consumers at the same time [4]. In Germany recycling costs are all covered by the producers according to regulations on a market-contract basis [5]. In Sweden fixed recycling fees were shifted from the consumers to the producers [6].

EPR is not a complete substitute to the market instrument. In other words, not all the EOL products need to be managed through EPR. Characteristics and amount of the potential values contained in and related costs for the treatment and management of EOL products principally affect the implementation of EPR programs and of the definite allocation of economic responsibilities. On the other hand, the less the EOL products are collected and recycled on the market, the more negative effects will be posed on the ecological environment by the EOL products. This paper will further examine the relationship among the recyclable profitability, market acceptability and negative effects on the ecological environment of the EOL products in order to further investigate the special scenarios for EPR implementations and crucial factors to be considered by the policy makers of waste management and also EPR.

2 Values components in EOL products recycling market

Treatment and recycling of EOL products mainly depend on the values retained in the products. As shown in figure 1, at the post-consumption stage products with economic valuable materials/components which can be reused or recycled will be collected and managed to get the reusable and recyclable materials/components going back to the supply chain. The residuals which cannot be used again, together with those EOL products having no economic reusing or recycling values at all, will be disposed in landfills. This is the case with most recycling systems where the recyclers and municipalities or the governments are the major parties involved.

As an interest-driven actor, the recyclers want to pursue the economic profits from recycling EOL products. Thus they have to pay all the costs during the whole recycling processes, including collection and transportation, disassembly and separation, repair or refurnishing for reuse, material recycling for recovery, marketing and other costs concerned. From the ecological point of view, the pursuing of the economic profits from EOL products would also be a sound way to improve the ecological performance of the discarded products. By reusing or recycling the valuable materials/components, the wastes for final disposal or landfill are minimized, the resources efficiencies are increased, and total ecological benefits are optimized. However, it is very hard to achieve a real Zero Emissions. In many cases, EOL products are far from the optimal reuse or recycling, or not being reused or recycled at all. In such circumstances, the external diseconomy occurs. And the whole society has to contain the negative value of the EOL products, or in other words has to pay for the ecological costs.

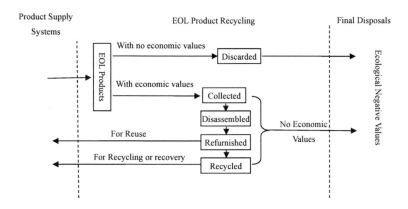

Figure 1: Recycling System of EOL Products

If we take a look from the economic point of view, recycling EOL products may cause three different parts of values, they are: (1) economic costs, including collection and transportation costs, disassembling costs, repair and refurnishing costs, recycling and recovery cost, marketing cost and other related managerial costs; (2) economic profits, referring to the net income from selling the reused and recycled materials and components, which is the economic profits of the recycling industries as well as the ecological values achieved through increasing resource efficiencies and reducing environmental pollutions; (3) ecological costs, or environmental pollution and destruction and natural resource abusing and wastes resulting from recycling residuals and also dumped EOL products which usually embodied as the landfill and final treatment payments. Among the above three values parts of EOL products recycling process, the ecological costs part is the real reason for the so-called external diseconomy or a kind of negative ecological performance which contained normally by the whole society. In this paper, we call it Negative Values (NV) of the EOL products. Next we will take a detailed look at the influence of the NV of EOL products on the final recycling extent and economic and ecological performances.

3 EPR scenarios and influential factors

3.1 Four different recycling scenarios for EPR

Due to different state of the products after consumption, different recycling technologies and system managements, and also different kinds of materials and components adopted during manufacturing, different EOL products have different recycling potentialities. Meanwhile, different ecological negative effects of the EOL products and different disposal requirements and charging standards will also lead to distinctive profit-making possibilities. Relative percentage of the above three values parts or the distinctive values structure of an EOL product would lead to different recycling situations.

Before forwarding a *Value Model for Recycling of EOL Product* as shown in figure 2, we set up the following hypothesis in order to facilitate our discussion. Firstly, the charges of landfill and final treatment for the wastes parts are calculated precisely based on the negative ecological values contained, which means the ecological costs are totally internalized into the disposal charges by the government. Secondly, the economic costs for disassembling, recycling and other processing of the EOL products before final disposal are taken as a fixed certain percentage, in order to strengthen the relative amount of negative ecological values and the profits gained on the markets. Thirdly, the market price is fixed so that the profit will be larger if the recyclable parts contained are larger, and vice versa.

The relative percentage of the economic profits vs. the ecological costs which will be partly paid as the landfill and final treatment fees will strongly affect the acceptability of the EOL products by the market. Therefore, we can classify four crucial values components scenarios as for the recycling of EOL products and for the implementation of EPR program. Let us name it the profitable scenario A, reluctant scenario B, refusable scenario C, and negative scenario D.

Under this framework we will try to explain why some of the EOL products are automatically managed by the market while some are refused. Also we will investigate when EPR instruments are needed as to dealing with the EOL products management problem. Besides, we will take a further look at the potential factors which we can control to achieve an ecological and economic sound solution of the EOL products management.

Profitable Scenario A: In this scenario, the discarded products have higher economic potentialities, which is higher than the negative values. Recycling such products will cause a positive cash flow in spite of the final ecological costs. See Figure 1(a). All the EOL products in this catalogue would be collected and recycled smoothly in the market, even though the governments strengthen the requirements for environmental management. For example, many EU members are going to limit the use of landfill, or increase the permitting standards for landfill or other disposal methods. Though such measures will increase the disposal costs covered by the recyclers, they can still make money by recycling EOL products in the profitable catalogue.

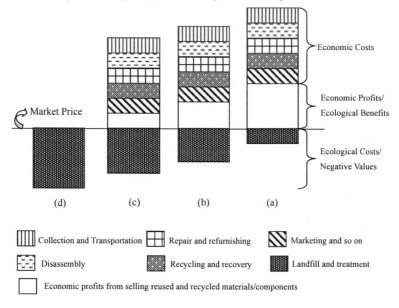

Figure 2:   Values Model for Recycling of EOL Product

Reluctant Scenario B: In this scenario, the discarded products have certain economic potentialities, which is almost equal to the negative values. See Figure 1(b). Recycling such products will cause a positive cash flow when the recyclers are charged less than to cover all the final ecological costs. If the governments strict their environmental management by increasing the disposal fees, the recyclers would probably earn nothing from recycling. But they could also have the possibility to get all the costs covered so as to keep their business running for some time. This is quite a marginal situation, in which both economic and environmental goals can be reached in case that strict environmental management and some basic economic incentive instruments such as tax preferable policies are provided.

Refusable Scenario C: In this scenario, the discarded products have much greater negative ecological effects than their profitable potentialities. See Figure 1(c). Recyclers were not able to keep running their business unless there is only a very small charge for them to cover the related environmental costs. However, it is obviously not so easy nowadays, especially in many developed countries governments are paying more and more attention to development sustainability. As a result driven by the market in such condition, recyclers would probably refuse to collect EOL products in this catalogue, and last owners would also choose to throw them away in the nature illegally in order to escape such higher disposal fees. Dilemmas between environmental protection and economic development appear to be a major problem facing most of the countries in this case.

Negative Scenario D: In this scenario, the EOL products do not passses many economic values to be recycled and are discarded directly to the final disposal stage as shown in figure 1. It would also be a result of *"Refusable Scenario C"*, when recyclers prefer to give up the economic values contained in the used products due to the comparative higher disposal costs. Consequently, no more used materials or components will be reused or recycled. And the whole discarded products turn out to be ecological burdens, see Figure 1(d). In such cases, it should be even harder to get either the recyclers or the last owners to pay for the negative values.

3.2 Influence of negative values on EPR scenarios

It should be interesting to see that all the costs and profits above dashed in figure 2 are related to the improvement of resource efficiency. The negative values or disposal costs dashed below are due to the increase of environmental pollutions. Thus recycling EOL products is indeed a combination of resource saving and environmental protection. Optimizing resource efficiency would be [necessarily] an effective way to decrease the total environmental burdens of EOL products, but not to eliminate it. When the negative values are as big as shown in scenario C and D, the environmental pollution caused by EOL products can not be solved efficiently in the market. There needs to be an innovative environmental management institution to have the ecological costs compensated [more efficiently] and the negative effects decreased more effectively. Here, an EPR program should be taken into force. EPR is also effective in situations of scenario B. Thus we call scenario B, C, and D the EPR Scenarios.

From the above analysis, we can see that implementing EPR is a way to compensate the extra ecological costs or negative values of EOL products while the EOL products would probably refused by the market and as a result a great ecological costs or negative effects on the natural environment would probably be induced. Factors which could affect the relative percentage of the negative values and the economic profits of EOL products should be seriously taken into account while considering and designing EPR. Specifically speaking, factors would affect the economic costs and profits by recycling EOL products and their monetary negative ecological values would also affect the decisions and design of EPR programs. Given a certain state of the EOL products, three crucial factors are being discussed in detail here.

(1) Disassembly and recycling technologies. In her doctoral dissertation the author proves that technology is the most influential factor in terms of promoting the producers to take the responsibilities of EOL products management based on empirical studies with Chinese automobile industry [3]. Technology is a decisive factor as for the economic costs and recycling feasibility of EOL products. Sometimes it will influence the expectations of the producers for future profitability in the condition that the producer is well aware of the situation of recycling technologies.

(2) Development of second-hand material market where economic profits realize at last. America is undoubtedly the most successful country in terms of making money from ELP vehicles. An important reason for her success is that there are quite supportive regulations to encourage the utilization of second-hand components and materials as the reuse and recycling of EOL vehicles. Such laws greatly motivate the development of such industries as disassembly, refurnishing, repair, remanufacturing and recycling. Development of this market will still depend on outside conditions, which means the policy makers could have the possibility to intervene into this factor.

(3) Accounting of negative values and charges for landfill and other final treatment measures. In the above, we prove the significance of landfill charges based on negative ecological values of EOL products for their final recycling scenarios. This will naturally lead to another question that: how to evaluate the negative ecological values of an EOL product and how to price the landfill. Levels of this factor differ a lot among different countries due to currency rate differences. This will not only hinder the motivation of recyclers to improve recycling technologies to decrease the recycling residues but also have the whole society contain a large amount of negative ecological costs made by the EOL products and residues as well as provide no scientific and obvious signals to the government to display the real ecological costs the society is paying for and the real ecological burdens the previous generations are leaving to next generations.

4 Policy implications for China

China is on its way to investigate the implementation of EPR. In its recently revised "Solid Waste pollution and Prevention laws", EPR has been forwarded as a new environmental management principle for the fist time. In the "Law on Circular Economy" which is still under discussion at present, EPR is formally regulated as an environmental management regulation. The automobile and electronic industries have been made the key industries in terms of implementing EPR. Detailed EPR policy and programs have also been discussed in these two focusing industries[3].

However, there are still some barriers China has to overcome first before the real implementation of EPR program. Actually the situation is very complicated for applying EPR in China. This paper will only take a focusing look at the key aspects according to the conclusions achieved above.

Firstly, the development of recycling technologies is far behind that of developed countries and the related management and processing are not normative enough, leading to lower recycling rates and serious second pollutions on site. Recycling profits are also much lower than in the developed countries even if the EOL products are with the same usage status.

Secondly, market for second-hand materials and components are developing very slowly. Normally the EOL products will only be treated in a very simple way, producing reused or recycled products with very low price in the market. No regulations encourage the development of second-hand market or ensure the quality of second-hand materials or components. Customers are also reluctant to accept such uninsured products.

Thirdly, charges for landfill and final disposal of the wastes are quite low in China. There are no accounting measures for evaluating the negative ecological costs of EOL products and wastes yet for the time being. So the current waste disposal charging standards fail to have the negative ecological costs covered by the economic systems themselves. Giant gaps of the landfill charges between China and other developed countries such as Japan, Germany and Canada greatly contribute to the fact that negative values of EOL products in China are already very large but no effective measures have been taken to deal with this. A main reason for this is that the existing negative values are not signalized in monetary form to show a clear picture to the whole society.

In general, some of the characteristics and factors discussed above could be controlled by the government, whereas some could not. Therefore, policy makers could take measures in such fields as the followings.

1) To promote E&D in the fields of reuse, remanufacturing and recycling.

2) To regulate strict standards for the reuse, remanufacturing and recycling of EOL products as well as to take measures to promote the development of second-hand materials and components markets.

3) To improve the current accounting system of the disposal costs of EOL products to incorporate the negative ecological costs to the charging standards.

4) To specify EPR policies based on different situations in different industries

Reference:

(1) Lindhqvist T. Extended Producer Responsibility in Cleaner Production——Policy Principle to Promote Environmental Improvements of Product Systems: [Doctoral Dissertation]. Sweden: Lund University, 2000.

(2) Naoko Tojo. Extended Producer Responsibility as a Driver for Design Change - Utopia or Reality? : [Doctoral Dissertation]. Sweden: Lund University, 2004.

(3) Zhao Yiping. Research on Implementing Environment and Functioning Mechanisms of Extended Producer Responsibility based on Chinese Automobile Industry. [Doctoral Dissertation]. P.R. C.: Dalian University of Technology, 2007.

(4) Yan Qiping. Recovery of Japanese EOL Vehicles and Implications for China. Journal of Chinese Waste steal, 2005, 5: 24-26.

(5) German Law Governing the Disposal of End-of-life Vehicles (End-of-life Vehicle Act - AltfahrzeugG) of 21. June 2002, (Federal Law Gazette I number 41 page 2199 of 28 June 2002) by the German Association of Motor Vehicle Importers.

(6) Forslind K. H. Implementing extended producer responsibility: the case of Sweden's car scrapping scheme. Journal of Cleaner Production, 2005, 13(6): 619-629.

2.

SUSTAINABLE DEVELOPMENT OF

INDUSTRIAL PARKS

2.1

Development and Trends of Eco Industrial Parks

The Development of Eco-industrial Parks Abroad and Its Enlightenment to China

Lu Chaojun, Qiao Qi*, Ouyang Zhaobin, Wan Nianqing, and Yao Yang

Center of Cleaner Production and Circular Economy Research,

Chinese Research Academy of Environmental Sciences, Beijing, 100012

Abstract: This paper reviews the development of Eco-industrial Parks (EIPs) abroad and analyzes their characteristics. Their successful experiences give us some enlightenment to implement EIPs in China. Moreover, there are some suggestions proposed on making further progress in EIPs' development in China, considering the environmental protection and the economic development, with aid of which it is hoped that the EIPs can grow healthily and sustainably in China.

Key words: Eco-industrial Park; Public Participation; Environmental Management System; Sustainable Development

Eco-industrial Park is the primary practice form of the theory of eco-industry, which seeks the balance of social economy, environment and human beings' demands. EIP integrates the advantages of cleaner production, eco-industry and comprehensive utilization of waste, in which resources are used properly, environment is sufficiently protected, and the harmonious progress of the regional industrial system and the eco-environment is achieved. In this way, the integrated benefit on environment, economy and society is generated. That is to say, EIP is one of the most important ways to establish the sustainable development in China.

1 The characteristics of EIPs abroad

1.1 The development of EIPs in the developed countries

The earliest Eco-industry Park is the Kalundborg industrial symbiosis system built in Denmark in the 1970s. It was built up with a power plant, an oil refinery, a pharmaceutical factory and a gypsum factory without interference by the government, imitating the ecology symbiosis principle.

*   Corresponding author: Q.Qiao, Tel.: +86 10 8491 5107, Fax:+86 10 84914626;
     E-mail: qiaoqi@craes.org.cn (Q. Qiao).
     First author: C.Lu, Postgraduate, Tel.: +86 10 8491 5646, Fax:+86 10 84914626;
     E-mail: lucj@craes.org.cn.

Said factories cooperated voluntarily and set up an eco-industrial chain in which one party used the wastes or the by-products generated during production of the other parties as its own production raw material. Therefore, not only the production cost is reduced, but also the waste produced in this area is decreased, coming to the "win-win" between the economy and the environment. However, there are some problems involved with the EIP. For instance, the factories depending on each other will have the negative effect that the change of the raw material or the production lines of the core factories, will affect the production efficiency of the collaboration enterprises. More seriously, once some factory goes out of business, the entire industrial chain will collapse.

There have been more than 20 EIPs in the United Sates since 1970s. The construction of the EIPs was dependant on the local conditions. Furthermore, various EIPs must set down the corresponding development goal according to the construction demands[1]. Four EIPs will be described as examples in the following. The first example is Fairfield EIP project located in Baltimore, Maryland. For the reason of further development of the local petroleum chemical industry, the original company is reformed by using the eco-industrial principle, and the matched factories, the ones dedicated on environmental protection technology, and the ones that use and circulate the wastes are absorbed. The next is the Cape Charles EIP in Virginia, which is developed by using the local natural and cultural resources. The Park takes the agriculture, the aquaculture and the travel industry and so on as the central industry, so as to develop modern industry technology aiming for demonstration of the economical and efficient use of the resources. The third example is the Chattanooga EIP in Tennessee, which pushed the reformation on "zero emission" by taking the recovery of nylon thread of DuPont Company as the core, as well as reused the wastes discharged by old industry companies to reduce the pollution of the park and promote the benefits. The last example is the Brownsville EIP located in Texas at the boundary of US and Mexico. There are five transportation modes profiting from its distinctive location superiority and developed traffic, which provides good conditions to build the "virtual" EIP.

About 40 EIPs have been built in Canada since 1995, involving various industrial combinations such as a combination of steam generator, paper mill and packaging industry; a combination of chemical industry, electricity generation, styrene, polyvinyl-chloride and biological fuels; a combination of electricity generation, steel, paper mill and flakeboard factory; a combination of thermal electric station, petroleum refining, factory and cement plant; and a combination of petroleum smelting, synthetic rubber factory, petroleum and chemical plant and steam power station. As the factories are gathered in the park, more and more factories are willing to take part in the EIP network, so that the EIPs will be provided with better balance and stable development.

In France, the government stipulates that the members of the EIP must obtain a Palme label which is specially designed for the EIPs by a France consultant firm. The local Palme group provides the factories in the EIP with the guidance handbooks related with the environmental management system, the analysis on life cycle and the resource conservation. These handbooks aim at the development of the entire Park, not that of a single enterprise[2].

By the way, EIPs are developing rapidly in Europe countries such as Netherlands, Sweden, Austria, U.K. and Italy. The Dutch EIP project is located in Rotterdam port. The Park is centered on self-determined participation of the factories and persists in the economic development and environmental protection thereof. The Park has 85 large or middle scale factories, with the goal to build an EIP mainly consisted of the petroleum or the petroleum chemical industry and the industry supported by them.

The EIPs in Japan keep ahead in the world and take the Vein Industry as the main body, which is the most distinguished characteristic thereof. Now there are 26 EIPs in Japan with the main content being reuse of the wastes generated in the parks[3]. Supported by the laws made by the Japanese government, the waste recycling industry is developing orderly and normatively. According to the related rules, the enterprises have taken the responsibility to dispose the industry wastes they generate. For instance, by the law for reuse of construction waste material, the project contractor is responsible to differentiate and decompose the buildings and reuse the waste material. The law for reuse of food says that the industry of manufacturing, processing or sale of foods should reuse the food wastes. If any factory has no capacity of disposing the waste, they must send the waste to the company that has the permitting license for disposing. In addition, the Japanese government actively supports the national EIP development by giving financial assistance to the factories which are engaging in reusing the wastes and advancing in technology. For example, the factories entering the EIP will obtain about one third to a half subsidy. Furthermore, some financial assistance will be given to the research and development of the eco-production and the technology related with "3R".

1.2 The EIPs in the developing countries

The EIPs are widely paid attention to in developing countries, like India, Philippine, Thailand and so on, whose continual development is hindered by the shortage of resources and the environmental pollution while their economies are booming now. Sustainable development becomes an ordinary demand. There are a lot of EIPs appearing in these countries, which always have some relations with the development of the economy of the EIP located.

An eco-industry network in southern Manila of Philippine which contains 5 industry parks services is one of the earliest eco-industry tests[4]. This program financed by UNDP (United Nation Development Programme) is called PRIME. It aims to construct a system for byproducts exchange and resources recycle between different parks by reforming said 5 industry parks ecologically and carry out research on the feasibilities of regional construction of resources recovery system and business incubator. There are 4 sub-projects in this program, i.e., 21st century agenda, eco-industry theory, environmental management system and environmental business, which coordinate and interact with each other. All the industry parks share the data and information obtained by the programme, and improve the inter-park byproducts exchange situation.

The EIP project has already become a national business in Thailand. At present, Thailand focuses on reconstructing all the national industry parks into EIPs under the leading of Thailand bureau of administration of Industry Park, which supports these EIPs and the enterprises both by finance and policy.

The Narodnaya EIP project in India, one of the biggest ones in the world, is constructing the EIP dominated by sugar manufacture like Guitang cooperation in China, which includes other industries, such as chemical industry, pharmaceutical industry, dye printing industry and food industry. The Narodnaya EIP is built into an eco-industry network program, seeking for a way to combine the prevention and treatment on pollution.

1.3 The characteristics of EIPs abroad

The Eco-industrial Parks abroad share some similarities in the aspect of development condition as follows:

(1) As lied in different development periods, developed countries who have finished their industrialization process build EIPs to cut down wastes produced by the Parks and to realize zero mission of the pollution in the Parks, while in developing countries the construction of the EIPs is closely related to economic development of the community to improve the utilization rate of resources and reduce the cost of production.

(2) It is easier to build an industrial chain and maximize the resources recycling in EIP construction by adjusting to local conditions, taking local pillar industry as the principle part and fully combining the enterprises in the Parks.

(3) The active participation of the enterprises in the Parks, the spontaneous cooperation among them and the policy guidance and financial support from the government promote healthy development of the EIPs.

(4) National environmental management system and some relevant laws and regulations strongly guarantee the construction of the EIPs.

(5) The concept of EIP wins popular support and the public participation is at a high level. The construction of EIPs is coordinated with the construction of the community.

(6) The construction of EIP network is attached importance to make the information be exchanged smoothly, sensitively inside and outside of the Parks and cover an extensive area, so as to help the enterprises know each other's production information and lay a sound foundation for them to establish cooperative relationship.

2 The problems existed in the development of the EIPs in China

The concept of EIP has been widely accepted and taken into practice in China. In August, 2001, the first demonstration site of EIP--Guangxi Guigang National EIP(Sugar Industry) was authorized by State Environmental Protection Administration(SEPA). After that the construction of the EIPs has been extended all over the country, such as Xinjiang Municipality, Inner Mongolia Municipality, Jiangsu Province, Shandong Province, Zhejiang Province, Liaoning Province, Guangdong Province, and Tianjin. The EIP demonstration sites cover the traditional industries like sugar industry, paper-making industry, chemical industry, cement industry, and metallurgy industry, and the high technology ones like electronic industry, environmental protection industry, automobile industry, and bio-chemical industry, as well as the construction of the vein industry Park. By March, 2007, the SEPA has demonstrated about 29 national EIPs. Besides, several EIPs are being planned. Compared with the developed countries, the construction of EIPs in China is lied at the initial stage, and the problems on environmental management, organization form and system guarantee are faced.

(1) Lack of sufficient cognition to the EIP project constructions. We started the EIP development just a decade ago. Various industrial parks titled as eco-industrial parks have sprouted at an unbelievable speed, many of which exist in name only. In contrast, the construction of EIP is developed gradually and orderly in the developed countries such as U.S. and Japan.

(2) Underdeveloped legal system and relative assurance systems. There are no interrelated laws and regulations, and no definite punishment mechanism for environmentally pollutional factories, as well as unreasonable charging policies for environments and resources, etc., which are not helpful in actively driving the factories within the Park to implement eco-industrial constructions.

(3) The barrier of communication between factories within the park. An eco-industrial park is constructed to provide a recycling system and an information system simulating a natural ecological system, where the byproducts of one factory can be used as the materials of the other, so as to maximize the usage of energies and minimize the waste discharge. However, if the factory holds back the information concerning the raw materials, products and wastes for the sake of protecting itself, the fundamental recycling function of the eco-industrial park can not be realized, thereby the above benefits can not be achieved.

3 The enlightenment derived from the construction of Eco-industrial

Parks abroad

According to the industrialization in developed countries, it is necessary to enhance the consciousness to protect the environment and improve the current industrial mode in the view of circular economy, so that the economy could be developed sustainably. Therefore, the EIP will benefit society, economy and environment simultaneously. What we should do is not to repeat the industrial development path in the developed countries, but to use their experience in this field for reference. This way, the environmental problems and lack of resources as well as the problems involving severe waste of energy and raw materials could be addressed radically in our country.

The economy of our country is booming now, accompanied with complex and various environmental problems, which involve different kinds of environmental pollution in developed countries at different developing periods, and the contamination gross that is always kept large. That is to say, the environment condition is severe. Therefore, the construction of ecological industry parks is proposed with higher requirements for benefiting both environment protection and economic development it proposes.

(1) Construct ecological industrial parks based on the local conditions. Currently, a large amount of industrial parks, economic developing-areas, and high-tech developing-zones have been established in many regions. Not only the development of industrial parks is promoted, but also investment can be saved largely, if we merely reintegrate the industries in the parks or introduce some new industries that can make symbiosis relationship with the old industries in the parks, and ecologically reconstruct the current industrial parks to develop ecological industry parks.

(2) Intensify the public education and publicity programmes, encourage public participation, and combine EIP construction with community development organically. Influenced by the conventional industry development mode, people in our country have weak consciousness about environmental protection. The ideas of ecological industry parks must go deep into people's heart, otherwise the construction of ecological industry parks can not obtain a long-term production. Not only are the education and training on environment for the industrial employees necessary, but also popularizing knowledge about environment and ecology for the community residents in the parks would be useful.

(3) Utilize government functions, and improve the environmental management system of parks. In order to develop ecological industry parks in our country, the support and drive from local government and the parks' management departments are needed. The government should create favourable investment environment, and intensify infrastructures for the parks, so that some powerful investors can be attracted. Meanwhile, in order to encourage industry to enter the parks,

the government should also develop various preferential policies on some aspects, e.g, finance, employee training, and tax. The development experience of ecological industry parks abroad indicates that guidance and support are especially necessary at the beginning of construction of ecological industry parks.

(4) Set up information platform. It involves the composition of noxious wastes and innocuous wastes, the flow of wastes, the destination of wastes, industrial manufacturing information, market development information, technology information, and talent information. It is helpful not only for the government to master the development of the parks, but also for industries to cooperate with each other and exchange the materials, energy among the parks, periphery areas and regions, which could make better use of resources in parks, reduce pollution discharge, and promote parks' continuous development.

(5) Speed the development of the EIPs. It is crucial to speed up the industry enterprises entering the parks, form the reasonable industrial layout and the perfect industrial chain, advance the cleaner production and the circular use, transform and promote the traditional industry, display the concentrating effect, and achieve the goal of integrated contamination treating and source control. Through the control of the gross contamination of the EIPs, we will achieve the goal of reducing energy consumption and waste discharge during the period of the Eleventh Five-Year Plan.

4 Conclusion

China is entering a new industrialization phase which consumes much more resources than ever before. Eco-industrial Park construction shall play a crucial part to achieve the following objectives in the 11th Five-Year Plan: By 2010, 1) The GDP per capita shall be doubled compared with 2000. 2) The energy consumption per GDP unit shall be decreased by 20% compared to 2005. 3) Industrial solid wastes utilization shall be increased to 60%. 4) The main gas pollution (SO_2 and COD) emission shall be decreased by 10%. We need to learn from the foreign countries their practices in doing Eco-industrial Parks, and try not to meet the same failures as them, and more importantly, we need to develop a deep understanding of the challenges and opportunities in front of us in combination with the current situation in China. We need to inspect and review the utilization efficiency of the resources overall from the respect of advocating systematic and integral Eco-industrial Parks construction, so as to inject more vitality and efficiency into the environmental management system and promote it to a new stage. Based on the Eco-industry, environmental planning fundamentals together with the uniqueness of each Eco-industrial Park, we will have a bright future to achieve sustainable economic development.

Reference

[1] Wen Yu, Zhong Shuhua. Characteristics and Trends of Eco-industrial Parks Development in the United States, Science and Technology Management Research, 2006,1:92-94.

[2] Xie Zhenhua. Theories and Practices of Eco-Industry. Beijing: Chinese Environmental Science Publishing House, 2002.

[3] Qiao Qi, Fu Zeqiang, Liu Jingyang, Wan Nianqing, Wang Jun, Yue Siyu. Assessment index system of eco-industrial parks. Beijing: Xinhua Publishing House, 2006.

[4] Lowe, Geng Yong. Bionomics and Eco-industrial Parks. Beijing: Chemical Industry Press, 2003.

The Existing Problems during Operation of Eco-industrial Symbiosis Network (ESN) and Research on Flexibility

Wu Chunguang[1], Cao Guozhi[2], and Qin Ying[2]

1. Dalian University of Technology, Liaoning Dalian 116024

2. Shandong University of Technology, Shandong Zibo 255049

Abstract: Eco-industry is a kind of industrial form, which imitates the way of physical recycling in the natural ecological process to perform an industrial ecological system. It is the optimum economical form of environmental protection, ecology and economical efficiency. But because of the complexity of its own path and operational mode, there are many problems, which have become the bottleneck of ESN, how to promote its flexibility is the premise of ESN .This paper tries to present the reasonable advices as references.

Keywords: eco-industry, flexibility, eco-industrial Park, network governance

0. Introduction

With the advancement of the industrialization process, the increasing environmental pollution and resource crises are threatening human's survival and social development. Facing the strong offensive brought by industrial civilization to eco-environment, human beings have to think over the modern modes of production, consumption behaviors, value and management methods we choose. During the past half century since the publication of 'Silent Spring' written by Rachel Carson in the 1950s, the strategy and activities of international environment protection have experienced a historical evaluation from end-of-pipe control, process control, industry regulation to function regulation, or from pollution preventing and abatement, clean production to eco-industry. As one of the newest industrial modes, an eco-industrial system aims to imitate the material cycle mode of natural eco-processes, and to make use of resources, energy sources and investment adequately in the material cycle of the inner production process of the system from raw material, semi-finished products, and waste materials to products. It is a regional system which is composed of enterprises, natural eco-system and a residential area. All of the elements work together and cooperate with the local government; exchange substances and energy designedly, share resources efficiently, consume resource and energy and produce as little waste as possible, and try their best to construct a relationship of sustainable development among economy, eco-environment and society.

The mode of eco-industrial parks can be recognized as the best operation mode according to present pollution reduction, utilization rate enhancement of resource and eco-environment protection. But because of the complexity of its forming way and structure, the mode of eco-industrial park is not perfect, there are still many problems and risks, and if there are not flexible management and operation, its development will be cumbered.

1. Problems and risks during operation of eco-industrial parks

1.1 Implicated organization structure and unfair benefits

The construction of eco-industry needs the recognition and participation of enterprises, government and society, as it will be infeasible without the support and cooperation of these actors. However, multi-participation can also cause so-called relationship risks, which are risks derived from the trust crisis caused by a lack of necessary communication between the different parties within an eco-industrial network. Two kinds of reasons can result in these risks, one is rational non-cooperation, and the other is irrational non-cooperation. Rational non-cooperation can be considered as rational non-cooperation behavior which is caused by the members' pursuit of profit-maximization, or can be named as opportunism behaviors. As for single enterprise, rational non-cooperation may be reasonable, but regarding the whole network, it is inefficient, and may be fatal for the whole network sometimes. However, unfortunately, each member of the network holds motivation of pursuing its own profit-maximization at the cost of the partners' benefits. Irrational non-cooperation does not relate to the pursuit of individual benefits, but it may be related to reasons such as information's distortion or magnification during transfer, incomplete information, or being unable to apply cooperation behaviors. In conclusion, according to the relationship of industrial network members, there are still some potential risks, the structure and relationship of organizations are instable essentially, and these instabilities may cause rupture to the network.

As to the relationship of symbiotic enterprises, the imbalance of competition status may destroy fair communication and cooperation of both sides. One of the main reasons of maintaining industrial symbiosis network is the balance of all participants' competition status, but with the exchange and update of techniques, resources and capabilities, the competition status of one enterprise may rise, while the others' may fall, the balance may be broken, the stronger one always regards the other ones as encumbrance, and this may result in communication and cooperation troubles between both sides, finally, the network may face risks of breaking up. Another cause of relationship risk is the asymmetry of benefits within the industrial network; this may cumber the fair cooperation.

According to the fair motivation theory, the one who suffers unfair treatment will try his best to get fairness back. Similarly, an enterprise which suffered unfair payoff mode may put on discontentment, and may even terminate the cooperation. At the same time, if one enterprise finds other collaborators got more benefits from the coalition, it may reduce the restriction for itself, or even take actions which are harmful to the network regardless its own benefits. Hence, in many cases, when they are evaluating the relations in the networks, co-workers think more of equity instead of efficiency [1].

Therefore, how to organize all participants to construct the park and to harmonize members' benefits and relations is the first problem to solve; secondly, how to construct the eco-industrial park under the market mechanism is also a problem of organization structure of the park. Thus, how to construct symbiotic cooperation relationship in the process of establishing industrial eco-industrial network is important for the construction of eco-industrial Parks in China.

1.2 Lack of technique and a complex production process

Above all, there are technical problems in the utilization of waste. The premises of changing waste to resources are as follows: there are certain waste and regeneration and separation techniques which are feasible economically, and just in this way, waste can be utilized by other enterprises. In the industrial symbiosis network, when one key enterprise changes its production mode, or an enterprise terminates its business, this may result in a shortage of some byproducts, and the whole exchange system may suffer serious interfering. If there are not enough standby suppliers, the symbiosis network may become very weak. Secondly, the instability of products may cause the relation of supply and demand to be instable, this may induce instability of waste product in terms of arranging and amounts, and this may also bring some difficulties to the cooperation. Furthermore, there are rigid problems in the network structure, though the resources and energy source may be utilized effectively through the network among members, within such symbiosis system, any member's activities may be influenced by others', once one enterprise operates abnormally, it may affect the stabilization and development of enterprises in the supply chain or even the whole park. For instance, the production process requires high quality materials, raw materials which are tally with standards of purity and quality are required, thus the room of using other byproducts becomes smaller. In Kalundborg, the Gyproc gypsum factory had to change its production process when it took in the byproducts from Asnaes power plant, because the temperature of gypsum from Asnaes power plant was different from the crude one. Though they improved the production process, and succeeded in the end, this still cost both of them a lot.

Therefore, there may be technical and economical problems when one enterprise thinks of changing the production process in order to make the byproducts be usable for others. The nutrition structure of an industrial system may not be less simple than that of a natural system. Hence, because of the restrictive factors of production process, it may result in the instability of industrial symbiosis network. For another instance, the production process of the purchasing enterprise (down-stream enterprise) which requires high quality raw materials may be difficult to accept the changes of characters or structures of materials from up-stream enterprise. And the case of Gyproc gypsum factory in Kalundborg eco-industrial Park of Denmark is a typical example. In 1995, Gyproc found there was a lot of vanadium in gypsum during its routine analysis, this metal harms a lot and may bring some metamorphic responses to some people. Through careful investigation, they found that the cause of the pollution was that Asnaes power plant tried out a very cheap material called Orimulsion from Venezuela. Orimulsion is a kind of oil from the Orinoco drainage area in Venezuela. The investigators found vanadium in the oil, and vanadium was also found in the gypsum. Asnaes power plant had no choice but to improve its equipments in order to prevent the accumulation of vanadium and pollution of gypsum was produced by the desulfurization equipment.

1.3 Differences of corporate culture and regional customs

Along with the enhancement of international economic integration, there may be enterprises from different countries in the same park. Therefore, participating enterprises in the industrial network are mixed, each of them having a different background of society, regulation and culture, and their own histories, values and beliefs. Thus, the concepts of managing, decision-making and behavior manners may differ as well. Consequently, during the cooperation of symbiosis network, corporate culture and regional diversity may inevitably bring all kinds of conflicts, which is an important factor affecting the safety of industrial symbiosis network.

Furthermore, there are some other obstacles when we establish eco-industrial parks in China, in which the first is the lack of law system, some of the present laws are not propitious to wastes' changing to resource, transportation and utilization after exchanging, and the programming and construction of the parks, and the irrationality of laws on environment and resource and charging policy may make enterprises be devoid of enthusiasm of constructing eco-industrial park; the second obstacle is the lack of management, nowadays, there are not enough excellent managers who can operate several independent enterprises, and harmonize the relationship of stockholders, community and government; the third is the lack of concepts, cycle and reuse of resources are important to eco-industry. Lots of consumers are doubtful of the products produced from reused resource, related departments seldom carry out such educating activities, and thus, the lack of concepts is also an important problem.

2 The analysis and research of flexibility

Flexibility can be described as the ability of adapting to uncertain conditions of systems. The analysis of flexibility is to calculate the magnitude of flexibility, examine how to attain flexibility, and how to increase flexibility of the system. As for eco-industrial system, because of the exchange of substance, energy and information, flexibility has become a bottleneck restricting the development of the system. Therefore, it is necessary to increase the flexibility of eco-industrial parks.

2.1 Establishing index systems of evaluating

The construction of eco-industrial parks takes a long time, hence the selection of techniques and products should be far-sighted. High-tech and products with good marketing prospects should be selected to ensure that the park will keep a certain market share over a certain period of time. The enterprises which are engaging in changing wastes into resource should be encouraged, and the new pollution and waste of resource caused by low level disposal should be avoided. Wastes are one kind of resource to a certain extent, but if it is a resource with low utilization level, it may produce new pollution after low level disposal, while enterprises with high-tech can utilize these wastes fully, and reduce the consumption a lot [2]. Starting with the demand of eco-industry itself, we should make great effort to research and develop green production process, techniques and its industrialization. The development of green industry and techniques offers technical support for the construction of eco-industry; it may make the network more perfect, and widen the constructing scope of eco-industry.

2.2 Unifying account and strengthening price flexibility

Under the condition of market economy, there are many uncertain factors. Beside what we have considered above, such as the issues of relationship of supply and demand, organization structure and benefits distribution, the other most important factor is the market price: For an enterprise in the park, it is crucial to keep prices stable to achieve the purpose of keeping the eco-industrial park stable. Taking the Lubei eco-industrial park as an example, it applies a unified account method to strengthen flexibility of the whole network. In this way, so as the whole network benefits, even if one enterprise operates to the bad, the industrial network may also work abnormally. And the range of the price accepted may be wider than that of each single account independently.

F can be used to measure the index of price flexibility. Generally, if $F \geq 1$, that means the system satisfies the requirement of flexibility. Or else, the fluctuation of price within the offered range may make an enterprise run in red deficit, and the flexibility is not eligible. For example, there are three enterprises which produce AP, vitriol and cement respectively constitute the whole industrial network of a chemical industry, the indexes of flexibility accounted in two cases of unified and independent are as follows: the unified result is F=2.35; the independent result is F=0.68. The bottleneck of independent account is the price of vitriol, when its price goes to 192 Yuan per ton, it reaches the bottom. If it goes on falling, the vitriol factory will run under deficit, and it will stop producing. Because of the conjunction of production process, the whole system may not work properly. While using unified account, we can pay more attention on the bottleneck which is prices of AP and cement, when they fall to certain numbers, they reach the boundary of flexibility. At this time, though the price of vitriol is also very low, the vitriol factory is in red deficit, as they can use all vitriol on production of AP by adjusting the output and do not sell them, so the price will not influence the benefits of the whole industrial network [3]. Thus, the ability of enduring price fluctuation of the unified account system is enhanced, and the flexibility of the system strengthened.

Furthermore, the improvement of production process and diversity can bring flexibility. According to industrial eco-system, because of the connection of members, improvement of one enterprise may influence others' operation. In this case, such improvement should be implemented under the unified arrangement of the industrial network. We can strengthen flexibility by improving management and producing process. Taking Lubei chemical industry as the example again, the flexibility of diversity derives from the diversity of supply of raw materials of products. In the industry chain of phosphate- sulphuric acid- cement (PSC), the materials include phosphorus gypsum from section of ammonium phosphate, gypsum without salt which is the byproduct from the saltern, and also the gypsum purchased from the outside [4]. Such diversity of materials sources makes the network more stable. The 'share and manage jointly' mode of Lubei eco-industrial park may also fit for other parks.

2.3 Strengthening flexibility by self-enforcing, interaction and integration and coordination Self- enforcing

In the operating process of eco-industrial parks, due to the complexity and uncertainty of practical economy and the existence of transaction costs, the contact   between special investigation and transaction is always imperfect , relying heavily on the third parties such as a court. These cannot avoid opportunism, especially in the network organization. We cannot take command and control means to manage and restrict like that inside of the enterprise, so participants' self-restrict is usually needed. Therefore, depending on the usual methods such as custom, honesty and credit standing to solve the issues of contracts, but not depend on the court directly and namely applying self-enforcing mechanism to maintain the industrial network is a feasible approach [5].

Interaction and integration

The interaction of industrial network modes is an endogenetic mechanism. The complex tasks during the operation of network organizations require groups to accomplish the process of offering products and services through interacting, and enforcing the connection of groups. The interaction mechanism shows the ability of individuals or teams to influence other participants directly or indirectly and the ability of responding to the environment. Through such mechanisms, the individuals or groups attain opportunities of getting resources from others and communicate latent resources or knowledge. On the other side, the interaction mechanism can not only promote the mutual understanding, but also strengthen the trust under guideline of fairness, avoid the misunderstanding during communication, so that there will be more opportunities to exchange ideas or resources. Interaction mechanism is important for enterprises with asymmetric resource because such enterprises can attain crucial resources by interaction to establish advantage.

As cooperation and competition are both the characteristics of the enterprises in eco-industrial parks, the interaction mechanism has two sides: the interaction of cooperation is based on credit, raising frequency of signing up and efficiency of executing contract depend on the relationship of cooperation. Such interaction will broaden the boundary of resource utilization, form common benefits and reduce the cost of coordinating enterprises. The interaction of competition provides conditions for collaborators to repeat the game of prisoner's dilemma. When groups repeat interaction expecting predictable future, the rational may increase the benefits and values, and reduce the latent opportunism [6].

During the operating of the network, because of the uncertainty of demands and complexity of tasks, the combination and cooperation of groups are realized by the integration of resources, and the interposition of society needs to integrate the mechanism to establish reliable and reciprocal relationships. Furthermore, the integration mechanism should serve new activities of network members, and efficient groups can be established by regrouping relational sequence freely. The integration mechanism derives from the emergence of new organizations and the stimulation of crucial adjustment of bilateral relations. As for enterprises, this is the result which is caused by two sides (consumers and suppliers). The integration mechanism forms new mode of multilateral negotiation by coordinating relationship of participants to consolidate the conjunct promises, release unilateral benefits of control to increase bilateral relationship investment, reduce uncertainty of market and dissymmetry of information, and form a new environment of competition and cooperation [7].

Coordination

In hierarchical organizations, orders, regulations and agreements are used to coordinate and protect the benefits of participants, especially the stockholders'. In the process of managing the industrial network, participants need to apply coordination to accomplish complex tasks under uncertain conditions. Though there are not conflicts, the participants should coordinate work allocation, activities and production. Essentially, the management of industrial network is not spontaneous, but based on efforts of coordinating consciously, if there are not such efforts, the network will break up. Therefore, coordination is the basic objective of the network management. [8]

3 Establishing the mechanism of trust and punishment of network enterprises

Trust is the basis of establishing industrial networks, and it is also the path of its healthy development. In order to avoid the speculative behaviors which may harm the benefits of coworkers, it is necessary to establish mechanisms of trust and punishment. Certainly, establishing mechanisms of credits and punishment is a complex social systems engineering. Not only systems, mechanism, organization, techniques, and laws should be considered and designed, but also the government, managers of parks, medium organizations and related department should cooperate closely.

Firstly, it is necessary to establish systems of associated credits, credits evaluation and credit information publication. Combinng the credit service, laws and government monitoring to form the mechanism of punishing behaviors of faith breaking, and make the mechanism to be the overawing power of preventing speculative behaviors.

Secondly, introducing and fostering multi-agency of credits in industrial parks to provide organizational guarantee for establishing credit system.

Thirdly, propagandizing and educating actively, to establish social public bases for credit environment and system. The local government and managers of parks should encourage enterprises to participate activities such as 'keeping faith' and 'fulfilling promises', take the enforcement of credit consciousness as the main content of forming culture of parks, in order to enhance the self-discipline ability, and stick up for benefits of the whole park.

4 Promoting the function of government's coordination and management

Because of the particularity of exchanging byproducts in industrial parks, governments play an irreplaceable role in the process of establishment and management of industrial symbiosis networks. The [functions of] coordination and management's fully playing is beneficial to its safety and stability.

Firstly, the government should take its advantage of policy and complete its service function to create a good environment for the industrial network. Secondly, the government should make 'third party 'play sufficient role, coordinate the contradictions and conflicts among the network members. Thirdly, it is important to constitute the whole blue print for the industrial network development, implement maintenance and management actively. From the angle of being propitious to the whole park, the government should program the development direction for the industrial symbiosis network, and put forward clear scheme.

Besides the above measures, strengthening the flexibility of eco-industrial systems is a long and meticulous work. In the running process of eco-industrial systems, there may be many uncertain factors from internal and external. The internal uncertain factors always include thermodynamics constants, physical property, delivery modulus etc, while materials supply, products demand and price, environmental conditions are included in the external factors. The existence of such uncertainties requires the process can still work normally, namely the process should possess flexibility, in despite of the uncertain values change within certain range. At present, there is much research on the uncertainty of systems. Generally speaking, the variables can be divided into two categories, one is the periodical alteration follows time, and the other is random alteration follows probability distribution. Generally, the former one can be solved by utilizing the multi-period model of issues; the latter one can be solved by establishing mathematical models.

In a word, flexible management and operation is the crucial point for smooth running of the eco-industrial symbiosis network, and its formulation needs a long time.

References

[1] Wang zhaohua. The research on eco-industrial symbiosis network[A]. Doctor Dissertation Research Paper[C]. 2003:34-37.

[2] Li Yourun, Xue Dongfeng et al. Suggestions on promoting construction of eco-industry in China[A]. Eco-industry: theory and application[C].2003:34-37.

[3] Li Yourun, Shen Jingzhu et al. System engineering of eco-industry[A]. Eco-industry: theory and application[C].2003:152-153.

[4] Feng Jiutian. The Lubei Eco-industrial Model of China[A]. Eco-industry: theory and application[C].2003:333-335.

[5] Burt,R.S. Structure Holes: the Social Structure of Competition[M]. Cambridge, MA: Harvard Business School Press, 1992, 231-235

[6] Axelord,R. The Evolution of Cooperation[M]. New York: Basic Book. 1994, 118-121

[7] Thoreli,H b. Networks: Between Markets and Hierarchies[J]. Strategic Management Journal, 1996,7: 37-51.

[8] Hu Shanying, Liang Rizhong et al. Analysis on industrial system of associated enterprises. Eco-industry: theory and application[C].2003:356-357.

Network and Resource Strategies in Industrial Parks
- The Cases of China and Germany -

Tiina Salonen

Faculty of Economics and Management, University of Leipzig

Introduction

In the past decades the number of industrial parks has grown rapidly worldwide. Due to the spatial proximity of companies industrial parks offer potential for the development of interorganizational resource exchanges in the form of product exchanges, by-product reuse, infrastructure sharing and joint services. The utilization of this potential enhances the resource efficiency and added value of companies and may reduce their environmental impacts. In the recent years efficient management of resources has gained in strategic importance particularly due to growing resource prices and ever stricter environmental laws and regulations.

The effective and efficient management of resources in industrial parks requires appropriate management concepts. The research and literature on the management of industrial parks in general is very limited. One of the few available theoretical approaches that deals with building of resource exchanges in the form of supply synergies, by-product reuse, infrastructure sharing and joint services at the industrial park and regional network levels is eco-industrial development (EID) (Chertow 2007, Liesegang & Sterr 2003). EID is part of the approach of industrial ecology that is based on the metaphor of an industrial system as a natural system. The aim in this kind of a system is to close the loops of material and energy flows and to act according to the norms for more sustainable economy such as connectedness, cooperation and interaction (Korhonen 2000, Ehrenfeld 2000, Cote & Cohen-Rosenthal 1998). Thus industrial ecology provides a vision and normative principles as well as operational tools for the integrated resources management in an industrial park. At the same time, however, it lacks a strategic management dimension. This deficiency has contributed to the failure of many hitherto initiatives.

This article provides insights into the integrated resources management in industrial parks, particularly the possible network and resource strategies as well as the strategy process. The work is based on different management theories and case studies carried out in German and Chinese industrial parks. The chapters below present extracts from the PhD thesis "Strategies, Structures and Processes for Network and Resources Management in Industrial Parks – The Cases of Germany and China" that has been submitted at the University of Leipzig in 2009.

Policy Framework for Integrated Resources Management

Public resource and environmental policies affect the development and operation of industrial parks directly or indirectly. Especially in Chinese industrial parks that are mostly governed by local authorities they have a considerable direct effect, whereas in Germany they primarily create the framework conditions for privately managed industrial parks. The primary policies that address integrated resources management[2] in the EU and China include:

EU Thematic Strategy on Sustainable Use of Natural Resources (in 2005)
China's Circular Economy Policy and Law on Circular Economy (launch in 2008)

The EU Thematic Strategy on Sustainable Use of Natural Resources (EU Resource Strategy) was finalized in 2005. It was as a response to the growing resource demand and prices that cause pressure on the competitiveness of the European economy and the environment. Moreover, it highlights the need for a concept that ensures secure resource supply to the EU also in the future, regardless of the geopolitical tensions and natural hazards in the areas where large reserves are located (EC 2005a).

An overall objective of the EU Resource Strategy is "to reduce the negative environmental impacts generated by the use of natural resources in a growing economy" (EC 2005a, p. 5). Since economic growth is the driver of resource use, the only sustainable way to reduce environmental impacts is to decouple resource use and associated environmental impacts from economic growth. This means utilizing natural resources more efficiently to protect the resource base and ecosystem services. To measure and evaluate progress on decoupling the following general indicators can be applied (EC 2005b):

Resource productivity: economic added value per unit of resource input [EUR/ kg]
Resource specific impact: environmental impacts (throughout the life cycle) per unit of resource use [impact/ kg]
Eco-efficiency: economic added value per unit of environmental impact [EUR/ impact]

---

[2] From the regional policy perspective integrated resources management is defined as a holistic approach that addresses the use of various natural resources by multiple stakeholders and the associated environmental impacts. Thereby it also assesses the linkages between different environmental media.

The concrete decoupling measures will be developed and carried out by the Member States and most relevant economic sectors that will be determined after the preliminary data collection (Hey 2006). Thus, the possible influence of the EU Resource Strategy on the operation of industrial parks in Germany will only be seen in the upcoming years.

The explosive economic growth during the past decade has tremendously changed the production and consumption patterns in China. This has led to resource scarcities and severe environmental degradation. To tackle these challenges China set a goal to reduce energy consumption per unit of GDP by 20 % and total emissions by 10 % by the end of 2010, whilst maintaining an average annual GDP growth of 7.5 %. Also an Action Program for Sustainable Development was released in 2007, in which a significant role is given to the Circular Economy Policy. The concept of circular economy has similarities with the EU Resource Strategy. It is defined as a sustainable growth model targeting both production and consumption and operating according to the 3R principle "reduce - reuse – recycle". Circular economy aims at increased resource productivity and eco-efficiency by closing the loops and a harmonious, "all-round well-being society" (Guomei 2007). Although often understood as a mere recycling approach, circular economy in the Chinese context can be viewed as a strategy for more comprehensive industrial restructuring, industrial policy reform and development of new technological, organizational and financial solutions.

Since the year 2000 circular economy strategies have been initiated at three levels in China:

Micro- or firm level in the form of cleaner production (Cleaner Production Promotion Law in 2003)

Meso- or industrial park level in the form of eco-industrial parks (EIPs) (Standards for Eco-industrial Parks in 2006)

Macro-level in the form of eco-cities and –provinces (Yuan et al. 2006).

At the industrial park level circular economy strategies can be developed within a company group or a development zone. By 2006 Chinese State Environmental Protection Administration had approved eight company groups and eight development zones with varying industrial focus as national demonstration EIPs. However, the Circular Economy Policy has encountered problems in the implementation at the local level.

Strategic Elements of Industrial Parks

Stakeholders

Based on the stakeholder approach an industrial park can be analyzed as a system of internal and external stakeholders that represent its actors. The stakeholder of an industrial park can be defined as an actor who can affect the achievement of the objectives of an industrial park due to his material or immaterial stake in it. At the same time he is also directly or indirectly affected by the operation of the park today or in the future.

Resource and network strategies in industrial parks should promote the competitiveness of the industrial park as a production location in the macro-environment or among other industrial parks and the competitiveness of tenant companies in their company industry environments. For the development of strategies and structures for network and resource management in industrial parks it is important to understand the interests and roles of different stakeholders. The key stakeholder in the network and resources management in an industrial park is the park management. It can be a public or private entity and takes the role of a park developer and operator. In addition, it acts as a regulatory authority in Chinese parks and as a service provider in German parks. Playmaker stakeholders of the park management are tenant companies. In general, they are customers of the park management and in privately managed industrial parks they may also be owners of the park management. In the public models in China the local government is a superordinate stakeholder of the park management, since the park management is a government agency. In privately managed German industrial parks the influence of the local government on the park operation is considered to be marginal. Further relevant stakeholders of industrial parks include surrounding communities and in China also research institutes.

Resources and Activities

From the resource based view an industrial park can be regarded as a pool of various tangible and intangible resources. They are used in a range of activities described by the value chain concept. Interfirm cooperation can be built to share resources such as raw materials, by-products, transport and environmental infrastructure, knowhow as well as management and information systems. Interfirm cooperation can also be established to jointly accomplish support and management activities such as purchasing, storage, emergency response, environmental management and communication. Industrial parks can also establish public-private cooperation with neighboring municipalities e.g. to share environmental infrastructure.

Sharing of resources and joint accomplishment of activities are coupled with material and information flows between the companies or the industrial park and neighboring municipalities. Thus, they constitute the content of interactions between the actors. In an industrial park the management of these interactions particularly at the interfaces between companies is of utmost importance. Potential for networking with regard to different resources and activities ranges from basic to more complex connections. The progression to more complex ones indicates the evolution of a more intense network and an improvement of the park performance.

Network

The coordination form of an industrial park is a network. Based on the type of actors involved the network layers in an industrial park can be divided into an interpersonal network, interfirm network and multistakeholder network. An interfirm network forms the primary network layer of an industrial park and is built between the tenant companies and between the tenant companies and private park management. The development of an interfirm network is strategic in the first phase. The park management closes contracts and establishes primary relationships with different tenant companies based on its own strategy and the needs of tenant companies. Furthermore, strategic cooperation in the beginning can be established between tenant companies as supplier-buyer-relationships. In the second phase the possibility of emergent network development through self-organization between the tenant companies rises based on established interorganizational communication structures and successful prior interactions. In industrial parks emergent development to utilize collective opportunities is also supported through relatively stable membership structures.

The overall coordination and controlling of the industrial park is organized hierarchically through the park management. The cooperation between tenant companies can be both hierarchic and heterarchic. Whereas the strategic cooperation between tenant companies having a supplier-buyer-relationship is often hierarchic, the emergent cooperation is usually coordinated heterarchicly. Network coordination mechanisms in an industrial park include contracts, trust, power and systemic interdependency, but their significance is context specific.

A multistakeholder network forms the secondary network layer of an industrial park. It is built between the tenant companies and public park management e.g. in China or between the industrial park and local government and communities e.g. in Germany. The types of multistakeholder networks that may exist in an industrial park include a policy network, a resource exchange network and a network focusing on joint accomplishment of business activities. A policy network focuses on the information exchange about a specific policy issue and serves the participative policy making and implementation among the public and private stakeholders. E.g. in Chinese industrial parks policy networks could serve the development of circular economy strategies at the park level.

Network and Resource Strategies in Industrial Parks

Integrated resources management in industrial parks aims at the enhancement of resource efficiency and added value of tenant companies and the park while reducing their environmental impacts. Its successful implementation requires appropriate network and resource strategies. The integration of the environmental goals and needs of tenant companies into the strategy development in industrial parks is of utmost importance as tenant companies are the playmaker stakeholders of the park management.

Network Strategies

The network strategies in industrial parks can be divided into overall network strategies and business network strategies. The overall network strategies comprise the following:

- overall park development strategy
- selection strategy for new tenant companies
- location factors
- strategies towards external stakeholders

The overall park development strategy depends on the management model. At present, the main management models in China include a public park authority and public private cooperation model and in Germany a private major user and private infrastructure company model.

In the public authority model the overall park development strategy follows the macroeconomic goals of the city government as the park management is its agency. The main objective in China has been the attraction of foreign investors, but lately also the Circular Economy Policy has gained in importance due to resource scarcities. In the public private cooperation model the park management consists of an administrative committee that is an agency of the city government and a development corporation owned by tenant companies. The development objectives include the attraction of foreign investors and enhancement of interfirm product exchanges and infrastructure sharing. In the major user model the park management is a unit of a large producing company. The primary objective of the development strategy is to enhance the value chain integration around the major user. The secondary objective is to build infrastructure integration. In the infrastructure company model the park management is a service provider owned by one or several tenant companies. Its main objective is to enhance interfirm infrastructure and service integration.

To realize its overall development strategy the park management follows a specific selection strategy for new tenant companies and offers favorable location factors. In the public authority model in China tenant companies are preferred from defined pillar industries and main selection criteria are the investment volume and company reputation. Due to the Circular Economy Policy investments of environmental industry are also encouraged. Typical location factors offered are preferential land and service prices as well as tax policies. In the public private cooperation and infrastructure company model the park management tends to focus on tenant companies from a single sector that need the offered services. In the public private cooperation model in China also foreign service providers are recruited with which the development corporation founds joint ventures. In the major user model preferred tenant companies include downstream companies and service providers that can be integrated into the existing value chain. Location factors offered in the public private cooperation and private models are the proximity to up- or downstream companies, lower fixed costs for infrastructure and possibility of concentrating on core competences.

In all models the park management also considers the environmental aspects of tenant companies in the recruitment phase. A standard criterion is the approval of an environmental impact assessment of a project. In Chinese parks the park management also sets caps to the resource consumption of companies. In German parks in turn the park management controls the compliance with umbrella licenses and companies must fulfill the environmental specifications of the land use plan and accept the Environment, Health and Safety (EH&S) objectives and principles of the park.

Strategies towards external stakeholders refer to common strategies of industrial parks for the management of relationships with e.g. authorities, neighboring communities or institutional investors. At present, the Chinese industrial parks have not developed any common stakeholder strategies. The German parks conduct e.g. an open dialogue with neighboring communities and communicate jointly with environmental authorities through the park management.

The goal of business network strategies of industrial parks is to realize economies of scope, scale and speed, flexibility and quality advantages as well as learning effects. In the area of integrated resources management they generally aim at the realization of economies of scale and shorter investment-to-market-times of tenant companies through sharing of environmental infrastructure. The German parks also offer one stop purchasing of environmental infrastructure services to reduce transaction costs and complexity. Quality advantages can be gained through the professionalization of utility and waste treatment operations by involving specialized providers. To achieve economies of scope and learning effects tenant companies can jointly accomplish EH&S activities through the park management and share associated knowledge.

Resource Strategies

The interorganizational environmental and resource strategies in industrial parks can be classified as follows (adapted from Gminder 2006, Meffert & Kirchgeorg 1998):

- secure
- credible
- efficient
- recycling
- innovative
- transformative

The strategy secure aims at risk reduction and control through implementation of risk management to ensure the legal security and societal legitimacy of business. In industrial parks the park management and tenant companies can jointly manage risks by establishing a common emergency response and safety management system.

The strategy credible supports the improvement of the image and reputation of a firm or a production site. The park management and tenant companies in industrial parks can jointly implement this strategy by demonstrating environmentally responsible behavior e.g. through the establishment of a common environmental management system, adoption of EIP labeling schemes, conducting active dialogue with neighboring communities and providing information to other interested third parties.

The strategy efficient endeavors the improvement of resource productivity and eco-efficiency of production. In industrial parks the park management and tenant companies can jointly increase their resource efficiency by utilizing each other's waste materials or energy, sharing environmental infrastructure and commonly conducting EH&S activities. Industrial parks can also share environmental infrastructure with neighboring municipalities through public private partnerships. The efficient strategies are often coupled with the realization of the business network strategies.

The recycling strategies aim at the reduction of the use of primary resources and associated environmental impacts by replacing them with secondary materials. Simultaneously less waste is generated and waste costs are reduced. In industrial parks the park management and tenant companies can implement recycling strategies by establishing material and energy exchanges. At present, the interfirm recycling strategies are mainly put in practice through specialized treatment companies.

The innovative strategy aims at the differentiation at the market through the development of innovative environmentally sound production structures, products or services. In industrial parks innovations can be jointly made by developing interfirm resource concepts and comprehensive service concepts e.g. for central EH&S services.

The strategy transformative endeavors the co-development of environmentally oriented transition of the economy and society. In Chinese industrial parks this strategy could be realized e.g. through tenant companies participating in the development of circular economy strategies for the park. The tenant companies of industrial parks in developing countries could also engage in corporate citizenship activities e.g. by cooperating with the local government and communities to assist in regional planning, to jointly develop infrastructure or to provide capacity building to local staff.

Important for the integrated resources management especially in Chinese industrial parks, where the target share for foreign investments of the total investments is ca. 70 %, are also the international corporate environmental strategies of foreign tenant companies. Multinational companies can either strive for the international standardization of environmental management by adopting a global company standard or national differentiation by adapting practices to local environmental requirements and competitive conditions (Epstein & Roy 1998). Different types of external stakeholder pressures tend to cause the standardization of varying aspects of environmental management (Christmann 2004). In Chinese parks the foreign tenant companies that have multinational customers and produce for export often adopt environmental standards beyond the local requirements.

Strategy Development in Industrial Parks

The strategies for integrated resources management in industrial parks can be divided into indirect and direct network strategies. Indirect network strategies are developed by the park management in a conventional strategy process, e.g. its service portfolio, or in cooperation with a tenant company, e.g. a customized service package, according to the resources and competences of the park management and the needs of a tenant company. Their implementation typically leads to collaboration of the park management and one tenant company. Indirect network strategies dominate the development of an interorganizational network in an industrial park in the first phase, but the extent of the development depends on the management model of a park.

Direct network strategies are developed in a network strategy process jointly by the park management and tenant companies according to the strengths and weaknesses of the industrial park and tenant companies and the opportunities and threats of their environments. Their implementation results in common actions of the park management and several tenant companies. The complexity of a joint strategy process can be reduced by limiting the participants to those companies relevant to the issue being addressed. In the process the network goals and strategies are determined in interorganizational workshops through consensus-oriented bargaining processes. Direct network strategies will drive the development of an interorganizational network in an industrial park especially in the second phase, if interorganizational communication structures are in place. The development of comprehensive integrated resources management in industrial parks requires network strategy processes. Until now the involvement of industry in Chinese parks using the public authority model has been insufficient.

References

Chertow, M. 2007. "Uncovering" industrial symbisos. Journal of Industrial Ecology, Vol. 11, No. 1, pp. 11-30.

Côtè, R. & Cohen-Rosenthal, E. 1998. Designing eco-industrial parks: a synthesis of some experiences. Journal of Cleaner Production, Vol. 6, No. 3-4, pp. 181-188.

Christmann, P. 2004. Multinational companies and the natural environment: Determinants of global environmental policy standardization. Academy of Management Journal, Vol. 47, No. 5, pp. 747-760.

Ehrenfeld, J. 2000. Industrial ecology: paradigm shift or normal science? The American Behavioral Scientist, Vol. 44, No. 2, pp. 229-244.

Epstein, M & Roy, M.-J. 1998. Managing corporate environmental performance: A multinational perspective. European Management Journal, Vol. 16, No. 3, pp. 284-296. European Commission. 2005a. Thematic strategy on the sustainable use of natural resources. 22 p.

European Commission. 2005a. Thematic strategy on the sustainable use of natural resources. 22 p.

European Commission. 2005b. Annexes to the thematic strategy on the sustainable use of natural resources. 16 p.

Gminder, C. 2006. Nachhaltigkeitsstrategien systemisch umsetzen. Wiesbaden, Deutscher Universitäts-Verlag. pp. 89-123.

Guomei, Z. 2007. MFA and circular economy development in China. UNEP International Workshop on Resource Efficiency and the Environment: Identifying Key Resource Flows, Tokyo, 25.09.2007.

Hey, C. 2006. Neuer Schwung für die zweite Halbzeit? Umweltstrategien der Europäischen Union. Politische Ökologie, Vol. 24, No. 102-103, pp. 18-22.

Korhonen, J. 2000. Industrial Ecosystem – Using the Material and Energy Flow Model of an Ecosystem in an Industrial Ecosystem. University of Jyväskylä, Disseratation. 130 p.

Liesegang, D. (Ed.) & Sterr, T. 2003. Industrielle Stoffkreislaufwirtschaft im regionalen Kontext. Betirebwirtschaftlich-ökologische und geographische Betrachtungen in Theorie und Praxis. Berlin, Springer. pp. 269-315.

Meffert, H. & Kirchgeorg, M. 1998. Marktorientiertes Umweltmanagement: Konzeption – Strategie – Implementierung mit Praxisfällen. Stuttgart, Schäffer-Poeschel. 830 p.

Yuan, Z., Bi, J. & Moriguichi, Y. 2006. The Circular Economy - A New Development Strategy in China. Journal of Industrial Ecology, Vol. 10, No. 1-2, pp. 4-8.

2.2

Improvement Management of Eco Industrial Parks

Discussion on the Application of Environmental Risk Analysis in Eco-industrial Parks Planning

Qiao Qi*, Li Yanping, and Yao Yang

(Chinese Research Academy of Environmental Sciences, Cleaner production and circular economy research center, Beijing, 100012)

Abstract: Environmental risk management is one of the important factors which influences the sustainable development of Eco-industrial Parks. This essay analyzes from three aspects including the transversal coupling industrial chain of Eco-industrial Parks, the vertical closing venous industry and the possible environmental risk caused by the complexity of the park after regional integration. Environmental risk management system in the Eco-industrial Park planning will be discussed from two aspects which are the planning target and planning contents of environmental risks by narrating the contents of identification, influence and solution to the environment risks in Eco-industrial Parks.

Keywords: Eco-industrial Park planning; environmental risk analysis; planning of environmental risk management system.

* Author: Qi QIAO (1963-), professor, research in theory of circular economy and ecological industry, cleaner production and eco-industrial parks construction planning .

1 Introduction

Eco-industrial Park is the major practice of eco-industrial theory. With the guidance of the circular economy idea, eco-industrial theory and cleaner production requirements, it has the advantages of transversal coupling, vertical closing, regional integration and structural flexibility. Since 2000, with the influence of the environmental pressure caused by the fast economical development and the international new thought of environmental protection, China has developed circular economy and established Eco-industrial Parks as an important measure to realize the target of regional sustainable development, and "win-win" results of economy and environment. As a new industrial development mode which is the solution to the structural pollution and regional pollution, adjustment of industry structure and industry distribution, and the establishment of an energy-saving and environment-friendly society, the Eco-industrial Park concept has gradually become one of the important approaches to increase the quality of various industrial parks and industrial concentration zones.

The main target of Eco-industrial Park planning is to establish a harmonious developing mode of sustainable development in industrial parks, reduce total volume and density of the contaminations in industrial parks or industrial concentration zones, reduce the regional environmental risks, improve the quality of the regional economy, realize the harmonization between industrial production and environmental protection, employ both energy-saving and pollution reduction measures, harmonize the development of ecology positive circulation and social economy, and establish the development mode of lower resource consumption and lower environmental damage. Eco-industrial Park planning mainly contains the analysis of the society, economy and environment situation in the park, the advantages (basis) and negative factors, the general frame design of the park construction, industrial system planning, environmental risk management system planning, pollution control, ecological environment construction planning, and the planning of key projects, analysis of the investment and profit and the guarantee system. Eco-industrial Park planning is the important basis as the guidance of the Eco-industrial Parks construction.

At present, State Environmental Protection Administration (SEPA) of China has authorized construction planning of 25 State demonstrative Eco-industrial Parks, of which some are smoothly developing, with wide foreground and good profit, while some are in stagnancy and facing failure. This result is caused by various reasons, with the increasingly serious situation of the environmental protection in China; Environmental risk has become an important factor for the sustainable development of industrial parks.

2 Environmental risk analysis of the Eco-industrial parks planning

2.1 Environmental risks in Eco-industrial parks

Environmental risk means the influence and damage to the mankind, society and ecology caused by "accidents" due to human activities or natural reasons. It can also be defined as following: the product of the probability of environmental accidents occurrence P (namely risk degree) and the environmental result caused by the accident C, and represented by risk value R, that is R (damage/unit time)=P(accident/unit time)*C(damage/accident).

Environmental risk analysis of Eco-industrial parks planning means identifying the potential environmental risks during industrial park development at the period of Eco-industrial parks planning and put forward a preventive planning by using the concept of life cycle assessment, integrated considering the connection between industries in the park and the regional relative environmental sectors.

Besides considering the common environmental risks of industrial parks, the important aspects which shall be considered aiming at the transversal coupling, vertical closing and regional integration of the industrial parks, including:

(1) Environmental risk of transversal coupling industrial chain

The ecological industrial chain between different enterprises and industries is the most outstanding character of transversal coupling of Eco-industrial Parks, and also an important measure of increasing integral resource use efficiency of the park and reducing the pollution emission. The circulation of materials and energy is the basis of stable development in Eco-industrial Parks, however, there are also environmental risks in the process of materials and energy reuse. In environmental risk analysis of Eco-industrial parks planning, the possible environmental risks in ecological industrial chain is the key point which shall be analyzed at the meantime of analyzing the general production system, storage system and relative infrastructures construction system, and reducing new environmental risks caused by Eco-industrial Parks construction to the largest extent.

(2) The environmental risks of vertical closing venous industries

Venous industries are generally using solid waste generated during production and consumption as the major materials, and transform them into resources and products which can be reused to reach the goal of energy-saving and environmental protection. It is an important process of connecting the regional material circulation in Eco-industrial parks, and also an important approach to realize the vertical closing of the park. Due to the character of this industry, it bears more serious environmental risks of contamination than others in its recycling, storage and selection processes, which makes the venous industry both the key process of reducing environment contamination and the point of frequently occurred environment risks. In environmental risks identification of Eco-industrial Parks planning, special attention shall be paid to the venous industry by increasing its environment-friendly degree and environment safety degree.

(3) The multiple environmental risks in regional integration systems

By integration of different industries, industry systems and natural systems, an Eco-industrial Park becomes a complex of nature, industry and society, which has its special ecological, systematical and sustainable ability. Compared with individual environmental risks, the result of environmental accidents in industrial parks will have wider influential aspects, deeper influential degree due to complexity after regional integration, and directly cause the increase of environmental risk value in the park.

2.2 The contents in environmental risk analysis of Eco-industrial parks planning

The contents in environmental risk analysis of Eco-industrial parks planning contains risks identification, influence analysis and solution analysis, and the major target is risks identification and its influence prediction, to form a pertinent planning by analyzing the environmental risk source, occurrence probability, control nature, project, measure etc. The key environmental risks factors considered in the planning are chemical leakage, fire disaster, explosion, accident of concentration pollution treatment facilities etc. The flow chart of the environmental risk analysis of Eco-industrial parks planning is depicted in figure 1.

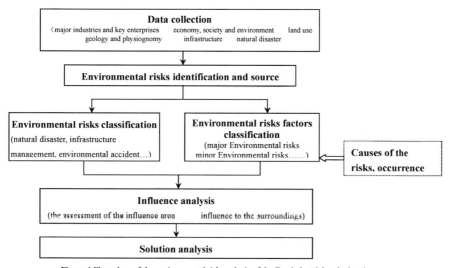

Figure 1 Flow chart of the environmental risk analysis of the Eco-industrial park planning

2.2.1 Identification and classification of environmental risks

The risk identification includes the production system, storage and transportation system and relative infrastructure system of the planned industrial park. The major methods for identification and assessment are statistical method and classified comparison to the risk sources and risk factors, including:

(1) Statistical analysis of various environmental risks in recent 5~10 years, select the environmental risks and factors with higher occurrence probability in the planned parks;

(2) For the raw materials, auxiliary materials, intermediate products and final products involved in the present enterprises in the park and the key projects, the toxic materials (extremely dangerous, highly dangerous), strong reaction or explosive, they shall be classified according to their physical

and chemical nature and toxicology nature, as well as their fatalness and toxicity combining with relative criterion, thus to filter the risks assessment factors.

(3) Identify the largest danger source and the most probable accident according to Appendix A.1 in *Guide on Environmental Risk Assessment* and *Identification of the larger danger source*, (GB 18218-2000).

The environmental risk types in the park can be classified into natural disaster, infrastructure management, contaminations accumulation, environmental accidents; the concretive expressive forms are fire disaster, explosion, the emission of toxic materials etc. The Analysis table of environmental risks includes the risk factors, causes, occurrence probability and damages of various environmental risks. (See table 1).

Table 1　The Analysis table of environmental risks

Type of the risk	Expressive form	Place	Occurrence probability	Damage degree
Natural disaster	emission of toxic materials	Chemical warehouse		Small
Infrastructure managements	The flooding of waste water with high concentration of heavy metal	Sewage treatment factory	Small	Medium
Natural disaster	Fire disaster	Electrical waste yard	Small	Small

2.2.2 Influence analysis

Influence analysis is the judgment and calculation to the occurring position, damage and damage nature, involving area, affecting degree of the risks, which includes the result assessment of accident and risk calculation.

Due to the uncertain implementation of the planning of the Eco-industrial parks, the correct calculation to the results of the environmental risks and environmental damages assessment is difficult. The influence analysis of environmental risks in Eco-industrial parks is generally more on a qualitative point of view to analyze the affecting area, the influence to the surrounded sensitive regions and the damage to human body and property. Generally, event tree analysis is used for qualitatively analyzing the probability distribution among the source, follow-up event and final result of the accident; Analyze the process of receptor safety risks caused by environment after emission of the pollution.

2.2.3 Solutions

There are mainly 3 types of approaches for controlling the environmental risks, which are reducing the environmental risks, transferring environmental risks and preventing the risks.

Reduce the environmental risks.

When the risks can not be prevented, then the risks can be reduced or eliminated by improving the technology, using the more advanced production techniques, technology and equipments, thus to increase the stability and safety of the production and raise the management level of risks, this method is the most widely applied at present. It is one of the effective ways to reduce environmental risks by comprehensively implementing cleaner production to form an energy-saving type increasing method of low investment, low energy consumption, low emission and high efficiency;

Transfer the environmental risks.

The position of the project can be changed or the surroundings of the project can be altered to make it tolerate environmental risks;

Prevent the environmental risks.

When the present technologies can not reach the goal of reducing the environmental risks, the projects which can cause larger environmental risks can be abandoned. This method is the radical measure to prevent the environmental method from happening. For example, the industrial policy of eliminating the out-of-date productivity unveiled by our country is the expressive form of this method.

In order to define the largest degree of the risk materiality and acceptability and find the solutions to minimize the risk, it must be achieved by analyzing the processes of environmental risk identification, classification and influence, and calculating the cost-benefit of releasing and assuming the risk The solutions' analysis include:

(1) Solutions for reducing the damages. According to the industries, position, environment character and the nature of the accident, analyze the measures to reduce the impact of the accident from the aspects of technology and management.

(2) Emergency measures analysis. It generally includes: the emergency organization and its responsibility; emergency facilities, equipments; emergency communication; result assessment; emergency monitoring; emergency safety, guard; medical emergency rescue; emergency evacuation measures; emergency report; emergency exercises etc...

3 The planning of environmental risk management system in Eco-industrial Parks

The planning of environmental risk management system is an important part of the planning of Eco-industrial Parks. On the basis of environmental risk analysis, by comprehensive assessment of the various environment factors in the park, an environmental risks distribution map can be made and solutions to the various risks can be worked out, and it can also provide a preventive and managing scientific basis of environmental risks during the construction of Eco-industrial Parks.

The long-term accumulation of the contamination in the surroundings has strong relations to the occurrence of outburst environmental events. In the planning of the environmental risks management, not only the prevention of the major risks shall be emphasized, the general environment affecting factors shall also be analyzed and relative preventive technologies and solutions shall also be brought forward.

3.1 Target of the planning

In the construction process of Eco-industrial parks, with the basic target of reducing environmental risks, with the major approach of controlling headstream and the whole process, special attention shall be paid to the major environmental factors, special risk factors and major risk sources, and influence the construction of the park by means of technological, economical, judicial, educational, policy and administrative, and bring forward the measures of reducing or controlling the environmental risks.

3.2 Contents of the planning

(1) Systematically analyze and evaluate the relativity among the potential environmental risks and its influence, the ecological capacity of the surrounding regions and ecological environmental sensitivity. Analyze the environmental sensitive point and the requirements of social economical development in the aspects of total volume and distribution by combining with environmental capacity, nature basis, the present social economical status and its future trend, then an integral conclusion can be made;

(2) Analyze the possible major accidents that may occur in the park according to the physical geography and social humanity character, predict the result of the toxic, flammable, explosive materials leakage (including ecological system and human body health), and put forward proper developing plan and management technology from the views of reducing, transferring and preventing the environmental risks.

(3) Enforce the construction of detecting ability of environmental risks. It includes clarifying the responsibilities and obligations of the government, enterprises and public in the environmental risk management, establishing organizations to prevent the environmental risks, developing publicized education of environmental risks prevention, improve the consciousness of the public regarding the risk prevention, and bring forward corresponding managing measure according to the specific risk factors in the park, such as chemical material storage and transportation.

(4) Define the warning level according to the risks level from the enterprises' and industrial park's view separately, make prevention, emergency treatment and emergency plan to the environmental accident with different levels and different types, bring forward the principle and requirements of the emergency information system, emergency management system, communication system, training system, warning system.

4 Conclusion

Eco-industrial Parks planning is an important part provided by the administrative department of the park for ecological industry construction as the technical and decision support. The characteristic of transversal coupling, vertical closing and regional integration of Eco-industrial Park has compensated the disadvantages of the former industrial parks which includes separate industries, linear production and dispersed system, while it also brings many environmental risks e.g. industries chain link broken caused by transversal coupling, venous industries secondary pollution caused by vertical closing, and the increase of environmental risks after regional integration which are increasingly obvious with the construction and development of the park. In order to reduce environmental risks to the largest extent by Eco-industrial Parks planning and realize economical, social and environmental benefit, due to the disadvantages of environmental risk management planning in Eco-industrial Parks, it is advised that the contents of environmental risk management planning system shall be supplemented and deepened, enforcing the research on the environmental risk management, the relative policy making, risk identification and preventive technology development etc.

References:

[1] Anil K. Gupta, Inakollu V.Suresh, Jyoti Misra, etc. Environmental rish mapping approach:rish minimization tool for development of industrial gowth centers in developing countries[J].Journal of Cleaner Production,2002(10):271-281.

[2] Guojun SONG, Zhong MA, Jing CHEN. Discussion on environmental risk and construction of management rules [J].Environmental pollution and control, 2006, 28(2): 100-103.

[3] SuoJun DU. Current Progress of Environmental Risk Assessment Research [J].Environmental Science And Management, 2006, 31(5):193-194.

[4] Gang DUAN, Xiaohai IIU. Probe into the Framework for Environmental Risk Assessment [J]. Sichuan environment, 2005, 24(4):59-63.

[5] Weihua ZENG. Research on Risk Forecasting and Model of Evaluating Environmental Pollution[J]. Journal of Disaster Prevention and Mitigation Engineering, 2004, 24(3):329-334.

[6] Jun BI, Jie YANG, Qiliang GAO. Study on Theory and Methodology of Regional Environmental Risk Zoning[M]Beijing: China environmental science press,Dec.2006.

[7] Zhenhua XIE.Theory and practice of Eco-industry [M].Beijing: China environmental science press,July,2002.

[8] GB18218-2000 *Guide on Environmental Risk Assessment* and *Identification of the larger danger source.*

The Research of Sector-integrated Eco-industrial parks Development Level

Liu Jingyang, Qiao Qi, Yao Yang, and Guo Yuwen

(Chinese Research Academy of Environmental Sciences, State Key Laboratory of Environmental Protection Industrial Ecology, Beijing 100012)

Abstract: Sector-integrated eco-industrial parks were defined. The development level was evaluation based on the evaluation index system of Sector-integrated eco-industrial parks in China. The research result of 11 sector-integrated eco-industrial parks shows that evaluation index system accelerated and orientated the development of eco-industrial parks. Focus on the problems of eco-industrial parks construction, advices on evaluation index system and eco-industrial parks construction were proposed.

Key words: Eco-industrial Parks; Evaluation index system; Sector-integrated Eco-industrial parks; Industrial zones.

1. Introduction

By the end of Dec. 2006, the State Environmental Protection Administration of China approved twenty-four National Demonstration Eco-industrial Parks, fifteen of which are comprehensive parks. Currently, there are fifty-three state-level high-tech industrial parks and forty-nine state-level economic and technological development zones, and also thousands of province-level, municipal-level and county-level industrial parks in China. Under the constraints of resources and environment, ecological transformation has become the eco-industrial park development model which can fundamentally ease the resource and environment pressure for "second pioneering". All types of industrial parks urgently need guidance for building Eco-industrial Parks. The content of the eco-industrial park can be simplified by an indicator system which attempts to simplify the complicated issue, in order to get popular indicators that help guiding the comprehensive and healthy development of the parks.

2. Sector-integrated Eco-industrial Parks Status Analyzing

The latest data (Dec. 2006) of eleven sector-integrated eco-industrial parks, all of which have been approved by the State Environmental Protection Administration of China, have been supplied. The results compared with the "Standards of Sector-integrated Eco-industrial Parks"[1] are shown in table 1.

(1) Economic Development Indicators

The industrial parks are at a rapid development stage in China, and the indicators of per capita industrial added value and increase rate of industrial added value have higher level in most parks.

(2) Material Flow Reduction and Material Recycling Indicators

From an overall perspective, this kind of indicator is less developed of the four indicator types standard demanded. There are four indicators with values below the standard value. No park reached the demanded rate of treated waste water reuse in the eleven investigated parks.

(3) Pollution Control Indicators

The indicators of discharge amount of COD per unit industrial added value and volume of SO_2 emissions per unit industrial value added are not satisfying while the other six indicators are completed very well in each park.

(4) Park Management Indicators

Most of parks have met the standard. The indicators of information platform perfectibility and public awareness rate to eco-industry are not satisfying in this kind of indicator.

3. Conclusions

The Ecological level of the parks is not satisfying in the whole. There is not even one park that meet the standards although the eleven parks have been approved by the State Environmental Protection Administration of China, which are superior in the country generally. So, the development of eco-industry parks still has a long way to go in China. Material Flow Reduction and Material Recycling Indicators are the key of eco-industry parks construction. The pollution control indicators and park management kind indicators will be improved significantly after years of building of key projects in accordance with the requirements of planning and will meet the demands gradually.

Table 1 Status Indicators of Eleven Sector-integrated Eco-industrial Parks

Items	Indicators	Unit	Index Value	Superior Indicators Parks		Junior Indicators Parks	
				Number	Target Range	Number	Target Range
Economic Development	Per Capita Industrial Added Value	¥ 10^4/per	≥15	9	¥ $30.9×10^4$~$15.1×10^4$/per	2	¥ $12.7×10^4$~$4.0×10^4$/per
	Increase Rate of Industrial Added Value	%	≥25	8	53.0%~25.9%	2	19.8%~7.2%
Material Flow Reduction and Material Recycling	Overall Energy Consumption Per Unit Industrial Added Value	tce/¥ 10^4	≤0.5	7	0.12~0.50tce/¥10^4	2	0.55~4.35 tce/¥10^4
	Volume of Fresh Water Consumption Per Unit Industrial Added Value	m^3/¥ 10^4	≤9	5	4.0~9.0m^3/¥10^4	5	8.7~30.38m^3/¥10^4
	Volume of Waste Water Produced Per Unit Industrial Added Value	t/¥ 10^4	≤8	5	3.02~7.97t/¥10^4	5	8.70~30.38t/¥10^4
	Volume of Solid Wastes Produced Per Unit Industrial Added Value	t/¥ 10^4	≤0.1	6	0.03~0.07t/¥10^4	2	0.2t/¥10^4
	Industrial Water Recycle Rate		≥75%	7	92%~80%	3	61%~56%
	Comprehensive Utilization Rate of Industrial Solid Wastes		≥85%	5	98%~86%	6	73%~45%
	Rate of Treated Waste Water Reuse		≥40%			7	35%~0
Pollution Control	Discharge Amount of COD Per Unit Industrial Added Value	kg/¥ 10^4	≤1	6	0.3~0.9Kg/¥10^4	3	1.3~9.5Kg/¥10^4
	Volume of SO_2 Emissions Per Unit Industrial Added Value	kg/¥ 10^4	≤1	5	0.6~0.8Kg/¥10^4	4	1.2~28.2Kg/¥10^4
	Hazardous Waste Treatment and Disposal Rates		100%	8	100%	2	95%~99%
	Living Waste Water Centralized Treatment Rate		≥70%	9	100%~75%		

	Living Garbage Sound Disposal Rate		100%	7	100%	2	78%~80%
	Waste Collection Systems		have	9	have	1	need perfect
	Waste Centralized Treatment and Disposal Facilities		have	9	have	1	need perfect
	Environmental Management Rules		perfect	9	perfect	1	better
Park Management	Information Platform Quality		100%	7	100%	3	25%~80%
	Park Preparation of Environmental Reports		1 issue/a	9	1 issue/year	1	0
	Public Satisfaction Rate With the Environment		≥90%	7	98%~90%	2	83%~87%
	Public Awareness Rate to Eco-industry		≥90%	6	97%~90%	3	0~20%

Note: the total number of superior and junior parks did not reach 11 for some parks can't offer statistical indicators

Reference:

1. Standard for Sector-integrate Eco-industrial Parks (On trial) HJ/T274-2006, State Environmental Protection Administration People's republic of China, 2 June 2006.

Eco-Industrial Network (EIN) Model, Analytical Methods

and Evaluation Index

Fu Zeqiang[1], Liu Jingyang[2], and Zhao Yiping[3]

(1: Academy of Environmental Science,

2: The State Key Laboratory of Environmental Protection and Eco-Industrial, Chinese Research Academy of Environmental Protection, Peking, 100012, P. R. C.,

3: School of Management, Dalian University of Technology, Dalian, 116024, P. R. C)

Abstract: Eco-industrial network (EIN) is the model for analyzing and stimulating the structures and functions of industrial ecological systems (IES). Based on the network analysis theories and methods, an EIN analysis model is set up, according to which indices are constructed for evaluating the structures and functions of the IES, including indicators such as associated degree, recycling path ratio, average path length and cycling index. Thus a basic analytical framework for EIN analysis is developed, which could be used as guidance for the evaluation, regulation and management of IES.

Keywords: Industrial Ecological System, Ecological Industrial Network, Index, Evaluation

The concept of industrial ecological system was forwarded by Frosch and Gallopoulos etc. at the end of the 1980s [1], according to which the framework of Industrial Ecology was formulated. The development of industrial ecological system (IES) is a long evolutionary process which is similar to the natural ecological system. IES is a given distribution of material, energy and information flow, highly depending on the supply of resources and services from biosphere [2]. A developed IES is composed of four major parts [3], that is resource exploiter, processor and manufacturer, product consumer and wastes collector and recycler.

At the early stage, industrial systems were basically inorganic processes, mainly built on physical and chemical mechanisms. The evolutionary processes of biological mechanisms with full recycling state were not completed yet. Every process was operated independently, and was only a simple superposition of some independent linear material flows[3]. Such systems were featured by dissipation and non-intensiveness, which means large resource extractions from the environment, while large of process wastes and discarded products after consumption discharges to the environment. Such mode leads not only to resource shortage and environmental pollution, but also to unfavorable economic performance. Therefore, reconstructing and upgrading the traditional industrial systems in ecological theories and methods of industry have become common awareness and strategies of all the countries over the world have started the ecological transitions and sustainable development of modern industries.

At present eco-industrial park (EIP) concept is a key mode to reconstruct traditional industrial systems eco-typically taken by various countries such as Denmark, the United States, Japan, Germany etc. As the most typical IES, EIP could be categorized into various types according to the components, functions and evolutionary features [4]. From the beginning of the 21st century, China has been promoting the planning of constructing EIP. Till now, 25 national demonstrating EIP have been initiated. During this process, related standards are also established by Chinese environmental managing departments [5]. Considering the needs of industrial ecological management, there are still deficiencies existing in present eco-industrial evaluating methods and indexes, especially the lack of those mechanisms indexes which can embody the structures and functions of IES. Based on theories and methods of network analysis, this paper sets up an analytical model of EIN, and then establishes indices for evaluating the structures and functions of the IES, including indicators such as associating degree, recycling path ratio, average path length and cycling index, in order to guide the evaluation, regulation and control and management of IES.

1 Basic Concepts

Although the differences of inner components (types and amounts of companies), symbiosis relationships and recycling mechanisms are different in structures and functions of distinctive systems, we can still stimulate

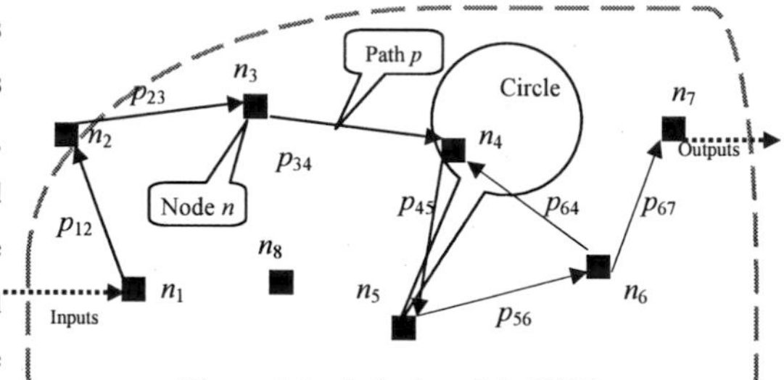

Figure 1 Analytical model of EIN

and analyze the structures and functions according to network theories and methods.

EIN is the model for analyzing and stimulating the structures and functions of IES. As shown in Figure 1, an EIN consists of nodes, paths and circles.

1.1 Nodes

The node stands for the structural unit with input and output within an EIN, which is in fact the companies' unit composing the IES. One node could get necessary nutrition from the outside nodes, and also discharge materials (product and by-product) to the outside or other nodes. The material flow structure for a single node is shown by figure 2.

According to the mutual relationships among the nodes demonstrated in the analytical model of EIN, the nodes could be categorized into isolated nodes, hanging nodes, and heavy nodes. Those nodes having no relationships with others could be called isolated nodes (such as n8 in figure 1. Those having relationships with only one node could be called hanging node, which could be further classified, in terms of the material flow direction, into input hanging node (see n1) and output hanging node (see n7). The former input materials from the system outside, and the later discharge materials out. Those having both relationships with the above two kinds of nodes could be called heavy nodes (see n2-n7), among which are the key nodes having the maximum relationships with the other node (see n4, n6). Companies represented by key nodes will be called "major industrial class" by industrial ecology.

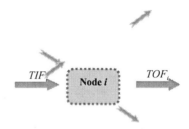

Figure 2   Material flow for single node

TIF_i stands for total input flow to node I, including those from other nodes and from outside the system;

TOF_j stands for total outputs flow from node I, including those to other nodes and from outside the system

1.2 Path

Connecting routes between nodes are called paths, which describes the material exchange relationships among nodes. The routes could be taken as having direction or not. Alternant appearances of the nodes and the paths consist of a chain (eco-industrial chain). There are at least two nodes and one path in a chain and lots of chains in a network system.

1.3 Circle

A circle also called recycling path or loop, which is a closed path composed of more than three mutually related nodes. Both the starting node and the ending one are the same. A recycling path consists of at least three nodes and three paths. Meanwhile, the number of the nodes and paths are always the same.

1.4 Associating relations

Association embodies the material exchange relationships between adjacent nodes in an EIN. If there are material exchange relationship between nodes, there will also be associating relations, and the associated time are 1. If no associating relations exist, then the associated times are 0.

Within an EIN, the material flow between nodes may be unilateral or bilateral, the former of which is called the unilateral association, and later the bilateral one. Both the unilateral and bilateral associations will be recorded once while counting the association times are counted. For example, node i provides raw materials to node j, their associating relation will be marked by Pij; at the same time, node j also provides raw materials to node i, then the associating relation will be marked by Pji, but the associating times will be only recorded once while counted.

Furthermore, associating relations could also be classified into forward associations and backward ones. Taking the node in figure 1 as an example, its association with n2 and n1 could be called forward association, with n3 would then be called backward association.

2 Analysis of EIN

Analysis of EIN contains the theories and methods for analyzing and evaluating the structural stability and material recycling efficiencies of an IES. Every company in an IES needs materials and energies supplied from outside or other companies inside. These materials and energies will be turned into specific products and by-products through physical and chemical processes, which will export to outside or other companies inside. It is normally agreed that complexity leads to stability, which means the more complex a system is, the more stable it is. The associating relations among various components inside determine the complexity and stability of the system. A developed IES is composed of the different companies with various material transforming functions to interplay. Inside such an IES, complex food chain and network structure are formed, which is represented as the chain, circle and their combination in an analytical model of EIN. As abstracting the mutual relations among companies in a actual IES into a EIN model, we can quantitatively evaluate the complexity and stability of its structure by analyzing such index as associating relations, path ratio etc.

Material flow analysis is (MFA) an important content for EIN analysis, which is widely applied in the analysis of natural ecological system. In 1970s, Finn analyzed the nutritious elements such as N and P in the Habbard break natural ecological system by adopting the method of MFA [6,7]. In MFA, the flowing features and transforming efficiencies of materials are generally described quantitatively by using the indices of average path length and cycling indicator CI [8,9], so as also to reflect the harmonic level between the system structures and functions. The bigger the average path length is, the more proved to be utilized by nodes before material leave this system. Cycling indicator reflects the material's recycling and utilization efficiencies level. The bigger the cycling indicator is, the higher the material utilization efficiency is.

2.1 Assessing index for structural complexity

2.1.1 Associating relation rate β

The Associating relation rate β is the index assessing the structural stability of an IES. Supposing the company number in an IES being n, the maximum possible associating times being M, and the actual associating times being m, the associating relation rate β could be calculated by the following formula:

$$\beta = \frac{m}{M} \qquad (1)$$

Thereinto: m——actual associating times;

$$M = \frac{1}{2}n(n-1)$$

M——maximum possible associating times,

The formula (1) could be further transformed into:

$$\beta = \frac{2m}{n(n-1)} \qquad (1')$$

Value of the associating relation rate β is ranging [0, 1]. It can tell from formula （1）:

(1) With the increase of β, the associating relations among companies will be enhanced, and stability of the IES will be improved.

(2) When $\beta=0$, the whole companies within the IES are all operating independently, there are no material exchange relations among companies at all. In this situation, all the nodes are isolated nodes and the EIN figure turns into a dispersed dots plot.

(3) When $\beta=1$, relations of material exchange are existing between any two of the compiles inside the system. In this situation, this system reaches its maximum association times.

(4) When the actual association times of an IES are n-1, and each company has only one connection with the other companies, all these companies in the system will consist on eco-industrial chain.

(5) When the actual association times of an IES are n, and each company has only one connection with the other companies,, then there is one and only one recycling path in the system.

2.1.2 Recycling path ratio η

This index is used to evaluate whether the recycling path amount in an IES reaches its maximum level. Generally, the bigger the recycling path ratio is, the more material circulating routes exist in the system, the longer the material rest on the system, the higher the transforming and unitizing efficiency is, and the less exchange of total input and output between the environments of system.

Supposing γ is the actual amount of the recycling path in an IES; θ is the maximum amount of a possible recycling path, then the recycling path ratio η could be calculated according to the following formula:

$$\eta = \frac{\gamma}{\theta} = \frac{2(m-n+1)}{(n-1)(n-2)} \qquad (2)$$

Supposing that there is no secondary inferior network in an EIN (networks without mutual connections), and recycling path exists, so the actual associated times are definitely larger than the minimum times n-1, then the amount of recycling path is equal to the difference of the actual associated times minus the minimum associated times, that is:

$$\gamma = m - n + 1 \qquad (3)$$

In any EIN with nodes over 2, the maximum possible associated times is, $\frac{1}{2}n(n-1)$ and the maximum amounts of recycling path in this EIN is equal to the difference between maximum possible association times the minimum association times:

$$\theta = \frac{1}{2}(n-1)(n-2) \qquad (4)$$

The value of recycling path ratio η is ranging [0, 1]. When $\eta=0$, there is no recycling path in the EIN; when $\eta=1$, the amount of recycling path within the EIN reaches its maximum.

2.2 Assessing index for functional level

2.2.1 Average path length PL

The Average path length is weighted average refers to denote average the total path length before being discharged from environment inputs of all nodes, which could be calculated by the following formula:

$$PL = \sum_{i=1}^{n} \left(\frac{z_{i0}}{\sum_{j=1}^{n} z_{j0}} ZPL_i \right), \quad i, j = 1, 2, \cdots, n \qquad (5)$$

Thereinto: zi0—Environment input of node i (including the inputs from outside and the other nodes)

$\sum_{j=1}^{n} z_{j0}$ —Total environment input of node i(same as the above)；

ZPLi—Length of inputting path of node i, could be calculated by:

$$ZPL_1 = e' \bullet M^O \tag{6}$$

In formula (6), e 'is n-dimension vector of heft 1, M^O refers to the first structural matrix of the material flow within an EIN, and could be calculated through:

$$M^O = (I - Q^O)^{-1} \tag{7}$$

In formula (7), I is the identity matrix, Q^O refers to the first diverting matrix of the material flow within an EIN. Q^O and M^O could be demonstrated as:

$$Q^O = \begin{bmatrix} P_{11} & P_{12} & \cdots & P_{1n} \\ P_{21} & P_{ij} & \cdots & P_{2n} \\ \vdots & \vdots & \vdots & \vdots \\ P_{n1} & P_{n2} & \cdots & P_{nn} \end{bmatrix} \tag{8}$$

$$M^O = \begin{bmatrix} m_{11} & m_{12} & \cdots & m_{1n} \\ m_{21} & m_{kl} & \cdots & m_{2n} \\ \vdots & \vdots & \vdots & \vdots \\ m_{n1} & m_{n2} & \cdots & m_{nn} \end{bmatrix} \tag{9}$$

Thereinto:

p_{ij}——Transition probability of material from node I to j, $P_{ij} = f_{ij}/TOF_i$; f_{ij} is the flowing amount from I to j； TOF_i is the total outputting amount of node i;

m_{kl}——Refers to the average times of materials diverting from node k to l.

2.2.2 Cycling indicators

In an EIN, supposing TCF being the total material recycling flow, and TST being the total substance flow (input and output), the cycling indicator CI could be calculated by following formula:

$$CI = \frac{TCF}{TST} = \sum_{k=1}^{n} W_k \bullet CE_k \tag{10}$$

Thereinto:

CE_k——Materials cycling efficiency on node k, $CE_k=(m_{kk}-1)/m_{kk}$, m_{kk} refers to average recycling times of materials on node k;

W_k——Refers to the weight of each node, could be replaced by the ratio of the material flow amount through node k divided by total material flow through the whole network, or $W_k=TOF_k/TST$。

3　Conclusions

Analysis of EIN includes the theories and methods for analyzing and evaluating the structural stability and material recycling efficiencies of an IES, and is an important content of industrial ecology systematic analysis. According to theories and methods of network analysis, an analytical model of EIN is set up, based on which several mechanism indices for evaluating the structures and functions of the IES are forwarded. Thus, basic analytical framework for EIN analysis is developed, which complements the content of systematic analysis of industrial ecology from the methodological level.

One thing that should be indicated is, that there may be more than two mutually independent network systems in a real IES. In order to simplify the analysis, this paper does not take the secondary inferior network into consideration when establishing each index, which should be improved in further research. Moreover, social and economic indices should be studied further in order to make a multi-dimensional evaluation to assess the development level of an IES.

References

[1] Frosch R.A. Gallopoulos N. E. Strategies for manufacturing. Scientific American. 261(3):144-152, 1989.

[2] Jelinski JW, etal. Proc Natl Acad Sci USA. 1992, 89:793.

[3] Snren Erkman. Vers Une Ecologie Industrielle[M]. Economic daily publishing company, 1999.

[4] T.E.Graedel, B.R.Allenby. Shi Hanze. Industrial Ecology (Second Version) [M]. Peking: Qinghua University publishing company, 2004.

[5] Chinese Environmental Protection Administration: http://www.zhb.gov.cn/

[6] Finn J T Measure of ecosystems structure and function derived from analysis flows J theory. Biol., 66: 363-380,1976.

[7] Finn J T. Flow analysis of models of Hubbark break ecosystem, Ecology, 61(a): 662-671, 1980.

[8] Han Boping. Cycling index of material flow in econetworks and their sensitivity analysis[J].Systems engineering-theory methodology applications, 1993, 2(4): 72-77.

[9] Jin Yong, Wei Fei. Circulation economy and ecolical industrial engineering[J]. Journal of Xi' an jiaotong university (social sciences), 2003, 23(4): 7-16.

Eco-efficiency of land-use - a decision support parameter for eco-industrial planning, real estate developments and brownfield re-use

Robert Hollaender

University of Leipzig, Faculty for Economics and Management Science,
Institute for Infrastructure and Resources Management,
Grimmaische Str. 12, 04109 Leipzig, Germany

Abstract: As a new decision support tool eco-efficiency of land-use is proposed. This parameter relates to the economic benefit of land-use decisions versus the entailed environmental aspects. Methodologies are proposed for the aggregation of the economic benefits and respective environmental aspects. The range of application is discussed and, as an illustration, some results of a first series of applications are presented.

1 Introduction

Economically prospering regions are often afflicted with a lack of developable areas. In this situation brown-field revitalisation offers a possibility to enhance the supply of available inner-city sites and reduce land reclamation in suburban areas. On the other hand, in less prospering regions there is an oversupply of potentially to be developed brownfield sites resulting in competing brown-field areas. Taking those competitive situations into account is an essential aspect of sustainable land use strategies. In this connection *eco-efficiency of land use* is proposed as a decision support tool for location decisions and land use options. Starting point of the method of *eco-efficiency of land use* is a definition of eco-efficiency for products and processes introduced in 1991 by the World Business Council for Sustainable Development (WBCSD) defining eco-efficiency as the ratio of the economic value of a product or process to its environmental influence [1]. The best ratio of environmental influence and economic value is theoretically characterized by extreme values in numerator and denominator. In practice however, the degree of freedom is limited and extreme values are not to be combined at will: the best ratio may be achieved by good environmental influences and very good economic efficiency or vice versa by very good environmental influences and good economic efficiency of the product. *Eco-efficiency of land use*

---

[1] Vgl. WBCSD (2000).

has, like eco-efficiency for products and processes, no absolute but a comparative character[2]. It is a measure that refers to the better choice of two or more alternative options when a decision has to be taken. *Eco-efficiency of land use* can be very well visualized. When plotting the aggregated environmental influences on the y-axis and the aggregated economic effects on the x-axis, you can easily read whether and in which way a deterioration or improvement regarding an existing land use option will happen, compare figure 1. For the diagram, economic and ecologic effects are standardized, so that the diagonal lines through quadrants B and C describe lines of the same eco-efficiency. The development option with the greatest positive distance to the diagonal line shows the best eco-efficiency. As shown in figure 1, this preferred alternative can also imply slightly negative environmental effects and very good economic effects and very good environmental influences and minor overall economic losses. In figure 1, compared to the intiial state the site development option in quadrant B has the highest eco-efficiency of the three development perspectives shown.

Figure 1: Eco-efficiency of land-use options

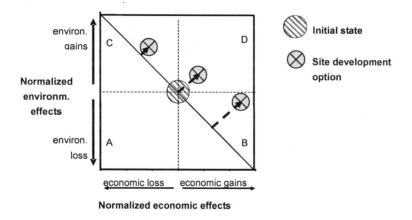

2   Aggregation of Environmental Effects

Aggregation of environmental effects is carried out in three steps. The first step assesses the influence of the construction of the development option on the environmental quality of the site: the environmental effects of a structural modification regarding soil, air, biotope quality, city and landscape quality. In the second step *environmental effects* of the day-to-day operation are evaluated. In the third step these two evaluations are combined to aggregated environmental effects.

---

[2] Vgl. Kicherer, A.(1999).

2.1 Assessment of the site specific environmental quality alterations

A qualitative, non-monetary evaluation process is required consisting of a measurement system and scaling, criteria as well as weighting factors of the criteria for the final aggregation. Generally, very diverse criteria can be used and evaluated on a rough or a fine scale. The following nine criteria are proposed in the urban and peri-urban context

- *Three soil criteria:* soil quality, soil structure, topography
- *Two (ground)water criteria :* ground water quality, ground water recharge
- *Two air criteria:* air exchange, cold and fresh air generation
- biotope quality
- urban and landscape quality

which are evaluated on a five-step scale in the range of "0" to "four" according to their quality. The aim of the method is a comparative evaluation. In this respect, for all options a comparable depth of information and a consistent procedure in the course of the decision making process are crucial.

Table 1: General description of the scale ranges

Value	Description
+4	- Subject of protection is in natural, unchanged condition
+3	- Proportion of the changed subject of protection: >0 – 25 % OR - Minor impairment or change of the subject of protection
+2	- Proportion of the changed subject of protection: >25 – 50 % OR - Major impairment or change of the subject of protection
+1	- Proportion of the changed subject of protection: >50 – 75 % OR - Strong impairment or change of the subject of protection OR - Suspected contaminated site existing
0	- Proportion of the changed subject of protection: >75 – 100 % OR - High contamination of the subject of protection requiring at least defense measures

The sum of all criteria equals the environmental quality of the location. The difference between various land use options is the difference of the environmental qualities "before" and „after" the change of the land use. A positive result represents an upgrading of the area due to changed land use options. A negative result, however, means a deterioration of the environmental state of the location due to changed land use options.

2.2 Assessment of environmental effects during the operational phase

Environmental effects are also caused by activities associated with the new land use option: emissions from the operation of facilities and effects by additionally triggered traffic volume. As the aim of the process is a comparative evaluation, largely identical effects can remain out of consideration. Along general lines, the same emissions can be expected from a production plant irrespective of using preferentially site A or B in one municipality. Differences in emissions evolve, however, if one site should be used for housing, a shopping centre or production facilities. Moreover, different effects on passenger and freight traffic will arise.

Here as well, the aim of a comparative evaluation facilitates the application of the method. It is sufficient to assess environmental effects of traffic in relation to other locations or other land use options and not in absolute terms[3]. Therefore, traffic impacts do not have to be completely recorded with all parameters. During the development of the method the environmental effects from changes in traffic demand, traffic volume and traffic flows were estimated and only recorded by one parameter as indicator, that is the additional carbon dioxide emission.

The location with the lower environmental impact caused by traffic is chosen as reference location. The basis of the method forms a grading in five steps, which has been introduced already in the course of the method of the ecologic quality of the location. Four steps are assigned to a relative increase of emissions resulting from street traffic by 25 % (Table 2).

Table 2: Qualitative scaling of the increase of emissions comparing location alternatives

Index	Increase of emissions due to additional traffic
+4	> 100%
+3	< 100%
+2	< 75%
+1	< 50%
0	< 25%

---

[3] A suitable estimation method was developed by the Wuppertal Institute on behalf of the Federal Environmental Agency, Wuppertal Institute for climate, environment and energy (ed.) (1999), applied here with a trip-end-model according to Schnabel et al. (1997) regarding travel demand modeling.

2.3 Aggregation

The weighting factor of a criterion documents the significance of the criterion in the evaluation method. Like in the eco-efficiency of products or processes it is possible to use different weighting factors to take different local and national preferences into account. Consistency of the evaluation within the evaluation and decision processes is crucial.

Weighting factors for the aggregation of site specific environmental quality alterations were derived for an urban German context. They place high emphasis on the change of "soil structure" and „soil quality" while „topography","ground water recharge", and "groundwater quality" as well as the air hygiene smoothing function to improve emission values of inner-cities by „fresh air generation" and „air exchange", all are assessed with an uniform lower weighting factor. Only the „biotope quality" is taken into account by an above average factor. „Urban and landscape quality" evaluates the cultural dimension, the visual connection of the land use with the surrounding environment and is weighted with a factor representing the average. The recommended weighting of the criteria is shown in table 3.

In order to aggregate the environmental quality of the location with the impact of the operational effects of the new or envisaged land use option only the additionally induced traffic effect is to be considered. For the latter a weighting factor that equals the average value of the other criteria is assigned, compare table 3.

Table 3: Evaluation criteria for the environmental condition of the locations and environmental traffic effects for Germany

Criterion	Weighting factor [%]
Soil structure	20
Soil quality	30
Topography	5
Groundwater recharge	5
Groundwater quality	5
Air exchange	5
Fresh air generation	5
Biotope quality	15
Urban and landscape quality	10
Environmental traffic effects	10

3.2 The Owner

The project development of a site depends on its status of planning legislation. Thus, a property site owner will realize a considerably lower sales price for a site not suitable for project development, than for a site, where legal planning requirements for a more significant land use option already exist. For brownfield-sites the last-mentioned preconditions are usually given. Generally, reintegration of brownfield-sites into the economic cycle is a difficult task for the owner, as characteristics of sites (property size and layout, neighbouring infrastructure facilities, noise, contamination risks, ...) make a land use option difficult to identify. Moreover, in many countries, the former owner is responsible for the remediation of brownfield-sites. This can lead to the situation where incurring remediation costs exceed sales revenues for property sites. Therefore, those property sites are often not being remediated despite payable real estate tax and current infrastructure follow-up costs and remain in the balance sheet of the proprietor. Revitalisation of such brownfield-sites is more likely when a positive value remains for the site after subtracting all revitalisation costs.

3.3 The Municipality

The municipality can generate income from land development actions. Income is generated as tax revenue in the course of real estate sales and the following land use. These taxes can be incurred within the land development project itself, additionally the project can generate an impulse for urbanization, leading to further municipal tax revenues. The municipality has a major influence on the possible added value for greenfield as well as brownfield developments. In Germany, the municipality can use the instrument of interim acquisition. This instrument can be applied to take part in the real estate market and to skim off profits, incurring when legal planning preconditions for a new development option are created by municipal decisions. Thus, the municipality can cover its costs to some extent. It can also use this instrument to avoid economic losses for the former owner or the investor and to facilitate land remediation, which would not take place otherwise. In the phase of interim acquisition the municipality can take advantage of state subsidies, which would not be available for private stakeholders.

3.4 Aggregation

The overall economic effects in this model are generated by summing up the single results of the stakeholders *land owner, investor* and *municipality*. For the aggregation a uniform time frame for all single results is examined. Costs and profits of the former owner, the investor and the municipality are estimated for the period observed. As a result, the overall profit of the land development measure is calculated. Figure 2 shows this method as schematic diagram. The former land owner has to bear operating costs as well as possibly reconstruction costs. When selling the property site he not only expects a compensation for the accrued costs, but also a profit. The municipality raises the total purchase price. Further costs may evolve, due to rededication, reconstruction, or marketing. The municipality as well will try to cover its costs at resale and realize a positive profit margin. The investor purchases the site at the price asked by the municipality and invests a further amount for the construction of a building. To put it simply, it is assumed that the investment for the building takes places immediately at a specific time T=0 and the construction phase is insignificantly short. Then, profit is realized directly from the utilization of the investment. Additionally, the municipality receives income from allocations and taxes. Figure 2 shows the aggregated effects from this process for the site. Thus, all revenues of stakeholders, which would continue to run as costs for one of the other stakeholders are omitted and the summary added value remains regardless of its distribution between the stakeholders.

It becomes apparent, that starting from an overall situation in deficit an overall benefit can be obtained after the investment. This overall benefit is used for calculating the eco-efficiency of land use in order to support decision making processes.

Figure 2: Aggregated economic results of site development options

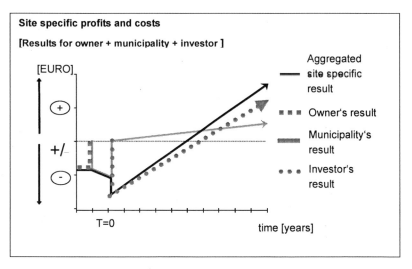

4 Case study

The development of the methodology was supported by the environmental research grant FKZ 205 77 252 of the Federal German Environment Agency. Test case applications were made possible through the support of the Federal German Railway System. The support of both institutions is gratefully acknowledged.[4]

Several brownfield-sites in East and West Germany in large and medium-sized towns were studied in the case study. The projected reuse options of brownfield-sites were compared with the respective greenfield land development options. For all studied brownfield-sites a higher eco-efficiency compared to the greenfield sites was identified. On the one hand, the environmental effects for all brownfield developments were rated better than those for the greenfield sites. In general, an improvement of environmental effects was achieved in the course of redeveloping the brownfield-sites, whereas a deterioration was noted for the greenfield sites, compare figure 3a and 3b. In three of five cases also the economic efficiency for the brownfield-site was higher than for greenfield site. In the remaining cases similar economic effects were assessed for brownfield-sites and greenfield sites.

Figure 3a

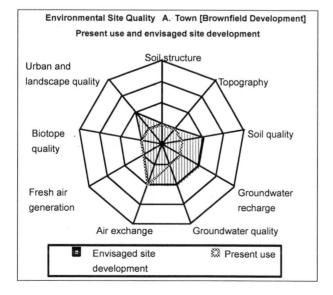

---

[4] Vgl. Holländer et al. (2009).

Figure 3b

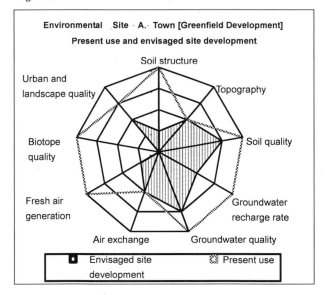

Environmental Site - A.- Town [Greenfield Development]
Present use and envisaged site development

Soil structure
Urban and landscape quality
Topography
Biotope quality
Soil quality
Fresh air generation
Groundwater recharge rate
Air exchange
Groundwater quality

■ Envisaged site development   ▧ Present use

5   Discussion and Conclusions

Eco-efficiency of land use is proposed as a decision support tool for location decisions and land use options. The definition is based on the definition of eco-efficiency for products and processes introduced in 1991 by the World Business Council for Sustainable Development (WBCSD) defining eco-efficiency as the ratio of the economic value of a product or process to its environmental influence. Accordingly, the land use option with the highest eco-efficiency identifies the best land use situation. The comparative character of this instrument is advantageous as it allows for a simple and very flexible application. However, applications can only be compared directly, when based on the same range of data with similar depth of information and when based on the same method to aggregate environmental and economic effects. In the case study the character of soil was weighted especially high. In compliance with national requirements or local traditions other weightings are also possible.

The introduced method allows for a comparison of development perspectives of several locations or different land use options. The method is especially suitable for prioritizing locations as regards urban development concepts. Prioritizing can be carried out as decision between two or more options. It can also be used in a relative examination by assessing the environmental quality for all sites of a pool of areas and by introducing a financial evaluation parameter. This can be the respective market value per unit area or the respective operating costs of the brownfield-sites to be compared. Often, revitalisation projects fail due to diverging economic interests of single stakeholders despite an overall economic gain, i.e., when high remediation costs prevent the development of a site or when only a social benefit can be expected. Characteristic examples are ecologic compensation or buffer sites, which, from the perspective of the former owner or the investor, cannot be turned into sufficiently profitable sites. A preceding assessment of the eco-efficiency of land use options for several sites facilitates the decision of the municipality, how revitalisation and project risks respectively and the achievable added value can be distributed to provide an advantageous land use situation („win-win-win") for all stakeholders involved.

Generally possible would be the application of other economic key figures, like the market value for several properties for an owner of a large supply of properties or the return on investment to be expected for an investor who has to make a choice between several properties. In these cases, a different eco-efficiency of land use would be calculated, which cannot be compared with the above mentioned overall municipal economic eco-efficiency, but which, however, presents a socially reasonable addition to the previous decision support parameters.

Within a research project, applications for several municipalities in Germany were tested with comprehensive model calculations. The test results lead to the assumption that a reasonable application of the eco-efficiency of land use can also be successful with less complex models as a decision support method for municipalities.

References

Doetsch, P.; Rüpke, A. (1998): Revitalisierung von Altstandorten versus Inanspruchnahme von Naturflächen: Gegenüberstellung der Flächenalternativen zur gewerblichen Nutzung durch qualitative, quantitative und monetäre Bewertung der gesellschaftlichen Potentiale und Effekte ; Forschungsbericht 203 40 119 im Auftrag des Umweltbundesamtes. Berlin: UBA Selbstverlag (UBA-Texte, 15/98)

Dosch, Fabian (2009): Mit Flächenbewertung zu einer Flächenkreislaufwirtschaft. 3. Refina Statusseminar. 23.03.2009. Online verfügbar unter http://www.refina-info.de/termine/ 2009-03-23-dosch.pdf, abgerufen am: 24.06.2009.

Holländer, R., Weidner, S., Schock, G., Pelzl, W., Lenk, T., Kühn, W., Thomas, E., Brandl, A., Kuhn, M., Rottmann, O., Winkler, C., Zacharias, G., Lautenschläger, S. (2009): "Nachhaltiges regionales Flächenressourcenmanagement am Beispiel von Brachflächen der Deutschen Bahn AG. Integration von Flächen in den Wirtschaftskreislauf." Abschlussbericht. Im Auftrag von Umweltbundesamt und Deutsche Bahn AG. Online verfügbar unter http://www.umweltdaten.de/publikationen/fpdf-l/3955.pdf, abgerufen am: 07.07.2010.

Kicherer, A.(1999): Ökoeffizienz zur Entscheidungsfindung bei Recyclingverfahren 8. Kunststoff-Recycling Kolloquium 06/1999 sowie Kicherer, A. (2001): Die Ökoeffizienz-Analyse der BASF. UmweltWirtschaftsForum, Heft 4, Seite 57-61

Schnabel, Werner.; Lohse, Dieter (1997): Grundlagen der Straßenverkehrstechnik und der Verkehrsplanung. Berlin, Wien, Zürich: Beuth.

World Business Council for Sustainable Development (WBCSD) [Hrsg.] (2000): Eco- Efficiency – Creating More Value With Less Impact, Genf.

Wuppertal Institut für Klima, Umwelt Energie GmbH (Hg.) (1999): Vergleichende Umweltbilanz - Umweltwirkungen von ausgewählten Einzelhandelsstandorten in Leipzig

The Spatial Layout and Development Strategy of Industrial Parks in Dalian

Wang Jinliang

(School of Management, Dalian University of Technology, Dalian, Liaoning 116024)

Abstract: This article is to discuss development features, major achievements and existing problems of industrial parks in Dalian in recent years. It is also to study the strategic architecture of spatial layout of industrial parks in Dalian and to put forward the spatial layout principle, structure and arrangement, layout mode and measure of future development of industrial parks in Dalian.

Keywords: Industrial Park, Layout Mode, Development Strategy, Dalian

1 Introduction

The industrial park is a way used to develop the economy and to improve the urban layout by some well-developed countries after World War II. The types of industrial parks are multiple and they are also differently referred to in various countries and regions. For example, the industrial park is referred to as "industrial site" in Japan, "industrial village" in Hong Kong, and "corporation district" in the UK. In general, the industrial park is a relatively independent area where a number of industrial corporations with different natures are incorporated and these relatively centralized industrial corporations are all under one superintendent unit or company that provides the corporations settled in the park with necessary fundamental facilities, services and administration.

It is indicated by the study of modern industrial economic theory that the formation of industrial parks is a product of social economic development to a certain extent, which is established to help enhancing the industrial centralization and satisfying the requirements for reasonable aggregation of urban development industries. It is an inevitable trend of social economy development [1-2]. In 1980, China began to establish economic districts in the regions along its southeast coast such as Shenzhen, Zhuhai, Shantou, Xiamen and Hainan. In 1984, China announced to establish development zones in 14 coastal harbor cities such as Dalian, Qinhuangdao, Tianjin, Yantai, Qingdao, Lianyungang, etc and set the first development zone in Dalian – Dailan Economic & Technical Development Zone. Multiple industrial parks such as hi-tech industrial development zone, export processing zone, tour & vacation zone, commerce development zone, industrial park, pioneering park, software park, environmental protection industry park and logistic industry park emerged thereafter in Dalian. The industrial parks in Dalian have made outstanding achievements by playing the roles of showcase, radiation, demonstration and have become a new place where the national economy has been growing, in the past years. However, the industrial parks in Dalian also have the problems such as large amount, dispersed distribution, unapparent scale merit, etc. Therefore, how to realize a reasonable layout of parks has become an issue to be urgently studied while the quick development of industrial parks in Dalian have been constantly facilitated.

2 Status of Industrial Park Construction in Dalian

2.1 History of Industrial Park Development in Dalian

In 1984, the first national development zone of the whole nation – Dalian Economic & Technical Development Zone was established in Dalian. Since then, the industrial park construction in Dalian began. So far, the industrial parks in Dalian have experienced the development for more than 20 years. The overall construction size has been constantly enlarged and the quality has been constantly improved. There have been 9 industrial parks at a provincial level or all over the city that have different types and functions, such as economic & technical development zone, hi-tech industrial park, free trade zone, etc. The history of industrial park establishment and development in Dalian can be divided into the following four phases in general [3-4] (Table 1).

2.2 Achievements of Industrial Park Development in Dalian

2.2.1 A New Growth for Economic Development

Economic growth speed averagely at over 20% has been achieved since the establishment of industrial parks in Dalian. It grows twice as quick as the growth in other areas during the same period. The profit and tax made by industrial parks in recent years have been up to one fourth of total financial income in the whole city. They have become the most important tax sources for local governments.

2.2.2 Promotion on Optimizing and Upgrading Industrial Structure in the Whole City

The industrial park directly raised the proportion of secondary sector industry in the whole city and the development of hi-tech industry drives extremely the technical modification of industrial corporations in the old city and the development of third sector industry, and speeds up upgrading the industrial structure. The industrial park prefers to choose introducing projects with high technical content, low energy consumption and low contamination that can drive upgrading local industries and enlarging employment. Dozens of projects over RMB 10 million have been introduced.

Table 1 Phases and Major Features of Industrial Park Development in Dalian

Phase	Time	Established parks	Major features
Beginning	1984 - 1991	Dalian Economic & Technical Development Zone, Hi-tech Industrial Park	It is the phase of industrial park construction and exploration with a focus on integrated economic & technical development zone. It is a basic development mode.
Quick developm ent	1992 – 1994	Free Trade Zone, National Tour & Vacation Zone at Golden Stone Beach, Jinzhou Economic Development Zone, seven city-level development zones, three county-level or district-level development zones	It focuses on city-level development zone. It is a high-speed development phase with the focus on comprehensive functionality. The development mode changes gradually from "building the nest to attract the bird" to "attracting the bird to build the nest".
Stable developm ent	1995 – 1999	Dalian Software Park, three county-level or district-level development zones, Tour Development Zone	Parks are stably developed. The construction in size and function presents the trend of stable development.
Overall improvem ent	2000 – presen t	More than 70 various parks	It focuses on the specialized construction of industrial parks. A lot of new industrial parks are established and the level of original parks is constantly enhanced.

2.2.3 Promotion of Town Formation

Along with the increasing expansion of the industrial park, agricultural land is converted into non-agricultural land, the countryside changes to an urban region, rural labor changes to non-rural labor, and rural population turns into non-rural population. In addition, the industrial park attracts a lot of employees from regions outside Dalian to relieve the employment pressure, which also increased the proportion of non-rural population in Dalian. The town formation level in Dalian has been raised from approximately 30% in 1980 to 63% in 2005.

2.2.4 Promotion of Adjustment to Urban Space Structure

With the opportunity to conduct urban construction and industrial structure adjustment in Dalian, the infrastructure construction and urban space expansion is made for the old city and the new urban area under construction as a whole. The industrial park becomes a primary carrier to conduct adjustment to urban space structure by space recombination inside the city. The industrial park effectively accommodates the corporations moving from the old city, and plays the important role of reasonably adjusting land structure in the old city and improving the city's overall functionality.

2.3 Problems Existing in Industrial Park Development in Dalian

2.3.1 Unreasonable Space Layout

It is seen from the entire city of Dalian that there is lack of integrated development planning in industrial park construction, and the industry type tends to be similar without effective combination of industrial layout and urban development space layout in the whole city. It is seen from the single parks that there is also a lack of planning in construction of some parks that have no advanced infrastructure, which is caused by the disordered function distribution in the park.

2.3.2 Administration System to Be Urgently Reformed

Currently there are certain problems existing in land use and administration in the industrial park: the one is that the procedures to deal with land certificate are so inconvenient that the progress of business and introduction of investment is even affected; and another one is that the land size in the park is not reasonable. Some parks have a too large planned area that causes a lot of land to be left unused, while some parks have a too small land size that limits the further aggregation of industries.

3 Strategic Architecture of Spatial Layout of Industrial Parks in Dalian

Dalian is a famous harbor city in North China, suitable for harbor proximity, opening advanced information and equipment fabrication industry. In 2003, the Chinese government made a strategic decision to build Dalian as an important international shipping center in Northeast Asia, and provided accordingly preferential policies in finance, tax, etc. This has brought a new opportunity for development of industrial parks in Dalian. The basic ideas are actively developing the industries such as shipping, shipbuilding, petrochemical, equipment fabrication, electronic information and software; making a quick and total development of modern service industry; and improving Dalian's role of opening, service, radiation and driving in regional economy development. Therefore, the current space layout must be adjusted for industrial park development in Dalian, to fully exert the industrial park's role in corporation rebuilding, displacement of external transfer land, hi-tech industry development and economic structure optimizing and upgrading.

3.1 Principle and Basis

3.1.1 Promotion of Industrial Group Development on Current Basis and around Advantageous Industries

So far, parts of the development zones have reached a certain size and formed a certain aggregation advantage. For example, the specialized industry zones such as water pump production in Fuzhoucheng, high-class clothes made in Yangshufang, exported furniture made in Huafeng, bearing production in Wawo, electronic manufacturing in Jinzhou Development Zone, etc have certain foundation advantages with which the expansion of park and industrial scale can be made on current basis.

3.1.2 Adaptation to Overall Target of Industrial Structure in Dalian

The affiliated services with dominant industries are a key choice of corporation development inside the development zone, and also a development direction of industrial park. Most of industrial parks in Dalian are developed under this mode, such as Mechanical Fabrication Affiliation Park in Jinzhou, Airport Logistic Park in Xinzhaizi and Petrochemical Industry Park in Shuangdao Gulf, as well as Biological Manufacturing Park, Free Trade Logistic Park, Software Park, etc inside the development zone that focus on high technology.

3.1.3 Saving Resources and Energy and Protecting the Environment

The special geological location, landform features, land resources, water resources in Dalian have become significant factors that influence industrial development in the city. The development of industries that consume large amounts of resources, such as papermaking, cement production, etc., is restricted.

3.1.4 Facilitating Reasonable Region Division of Industries in the Industrial Park

The inter-region industrial and economic cooperation is an external impetus to industrial development at market economy conditions, which can exert the government's macro direction, intensify market reformation, cancel the current system of inside and outside division, region division and structure division in the industrial park and make mutual economic benefit.

3.2 Structure and Arrangement

3.2.1 Industrial Parks Preferred to Be Supported and Developed

They include Dalian Economic & Technical Development Zone, Dalian Free Trade Zone, Dalian Hi-tech Industry Park, Dalian Export Processing Zone, etc. The advantages of this type of parks are that they are located in the city area, have apparent regional advantages and are highly competitive in infrastructure investment level, industrial foundation construction and corresponding administration system. The problem is that the construction land is subject to factors such as urban planning, land price, etc. For this type of parks, the development of hi-tech industries with high technical content, low resource consumption, less environment contamination or human resource advantages that are fully exerted is encouraged.

3.2.2 Industrial Parks to Be Significantly Supported and Developed

They include Harbor Industry Zone in Changxing Island, Yingchengzi Industrial Park, Huayuankou Industrial Zone, Gulf Industrial Zone, Wafangdian Industrial Zone, Songmu Island Chemical Industry Zone, Sanshilipu Harbor Industry Zone, Dengshahe Harbor Industry Zone, Lvshun Economic Development Zone, Piyang Industrial Zone, Shuangdao Gulf Petrochemical Industry Park, etc. Most of these parks are adjacent to the sea and their industry scale and aggregation advantages have been formed. The infrastructure is complete. The geological location is suitable. The land use conditions are relatively easy. They are the key areas to accept industrial structure adjustment and corporation moving and rebuilding in the city center, and also the regions for development of heavy and chemical industries and equipment fabrication industry adjacent to the city harbor.

3.2.3 Industrial Parks to Be Generally Supported and Developed

They include Paotai Economic Development Zone, Dalian Xianyu Gulf Province-Level Tour Vacation Zone, Jinzhou Economic Development Zone (Jinzhou New Industry Zone) and Pulandian Economic Development Zone. These parks are located around key small towns and have specialized features and adequate sizes. The specialized division in the park is promoted. They will be developed in combination with corporation affiliation and multiple operations.

3.2.4 Industrial Parks to Be Self-developed

They include Shihe Economic Development Zone, Ershilipu Economic Development Zone and Xianrendong Economic Development Zone. The town economy conditions on which these parks are based are ordinary, and their infrastructure is not complete. They can seek for future development through specialized regional cooperation.

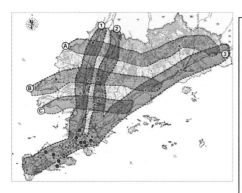

(1) Fabrication Industry Zone along Shenyang—Dalian Freeway;
(2) Fabrication Industry Zone along Harbin—Dalian Railway;
(3) Harbor Proximity Industry and Agricultural Product and Byproduct Processing Zone along Dandong— Dalian Boulevard
(A) Dispersed Fabrication Industry Zone along Yongning—Qingduizi Line
(B) Ordinarily Centralized Fabrication Industry Zone along Chengzitan—Bachagou line
(C) Highly Centralized Fabrication Industry Zone along Haicheng—Pikou Line

Figure1 Sketch of Space Layout Mode of Industrial Parks in Dalian

3.3 Layout Mode

A part of the industrial parks is integrated and rebuilt in accordance with spatial layout features and future development types of industrial parks and in combination with regional features of Dalian, under the "point-axis" space structure system, to form the space layout mode of "three-longitudinal and three-latitudinal" (see Fig. 1).

3.3.1 Three-Longitudinal Mode

The three-longitudinal layout mode of space layout of industrial parks in Dalian covers the fabrication industry zone along Shenyang—Dalian Freeway, the fabrication industry zone along Harbin—Dalian Railway and the harbor proximity industry and agricultural product and byproduct processing zone along Dandong—Dalian Boulevard. Shenyang—Dalian Freeway (fabrication industry zone) includes Jinzhou Economic Development Zone, Paotai Economic Development Zone and Sanshilipu Economic Development Zone. Harbin—Dalian Railway (fabrication industry zone) includes Pulandian Economic Development Zone, Gulf Industry Zone and Wafangdian Industrial Park. Dandong—Dalian Boulevard (harbor proximity industry and agricultural product and byproduct processing zone) includes Dalian Hi-tech Industry Park, Dalian Software Park,

Dalian Economic and Technical Development Zone, Dalian Export Processing Zone, Dalian Free Trade Zone, Lvshun Economic Development Zone, Huayuankou Industrial Zone, Yingchengzi Industrial Park, Piyang Industrial Park.

3.3.2 Three-Latitudinal Mode

The three-latitudinal layout mode of space layout of industrial parks in Dalian covers the dispersed fabrication industry zone along Yongning—Qingduizi Line, the ordinarily centralized fabrication industry zone along Chengzitan—Bachagou line and the highly centralized fabrication industry zone along Haicheng—Pikou Line. Yongning—Qingduizi Line (dispersed fabrication industry zone) includes Yongning Economic Development Zone, Xianrendong Economic Development Zone and Qingdui Economic Development Zone. Chengzitan—Bachagou line (ordinarily centralized fabrication industry zone) includes Changxing Island Harbor Industry Zone, Wafangdian Industrial Park and Huayuankou Industrial Park. Haicheng—Pikou Line (highly centralized fabrication industry zone) includes Paotai Economic Development Zone, Gulf Industry Zone, Pulandian Economic Development Zone and Piyang Industrial Zone.

4 Conclusion

The development of industrial groups has corresponding advantages. First, the space aggregation can realize the share of various resources (including infrastructure, labor, information and knowledge), the intensification of competitiveness in saving corporation production costs, forming groups, and the promotion of regional economic development. Second, the high confidence and cooperation culture established inside the industrial group and the economic association and social relationship established thereon form a specially combined industry atmosphere to spread knowledge, information and culture. Such exchange and share drives in turn the independent innovation in the region to produce more competitiveness. Currently, the city government in Dalian can support the healthy development of industrial groups in the following three respects: one is to build up the external environment for regional cooperation with the guarantee to establish consolidated market competition rules; and another one is to provide powerful support to regional cooperation with the integration of infrastructure construction; and the third one is to create essential material conditions for inter-regional cooperation by mutual industry advantage supplementation. Meanwhile, the system innovation must be tried such as establishment of policies for land supply, tax, finance, etc that break through "administrative division". A part of industrial parks are actively integrated and rebuilt to form the industrial group and therefore to improve the mass benefit and achieve the goal from industrial aggregation through mass benefit to economic growth.

The small and medium sized enterprise parks in a county region are mostly established on low cost, low price and even cheap labor price. The bad competition between corporations is inevitable. Moreover, the administrative division results in lack of regional coordination and cooperation between local governments and causes regional structures becoming involving low-level repeated constructions. In face of economic globalization and knowledge economy, these low-level industrial parks would be non-competitive if innovations for them are not made. There is a good town-regional economic foundation in Dalian. It is a shortcut to innovate and rebuild small and intermediate enterprise parks with advantages on natural resources, knowledge resources and industrial foundation, through directing the large leading corporations to be split and the small and intermediate corporations to be aggregated around different industrial chains.

Industrial park development is substantially the market corporation's self organization. However, almost all successful industrial parks in the world are directly related to the effective supply of public products and the powerful support of public policies from the government. The government's direction on industrial parks is primarily oriented at industrial direction, public service and market supervision. Currently the government can achieve the goal of reasonably adjusting the industrial structure in the whole city through the direction on industry choice and the market supervision. Industrial structure optimizing and adjustment is realized through: preferring to develop information industry; actively developing hi-tech industries that drive the economy to grow strongly; modifying traditional industries with advanced technology and greatly rebuilding equipment fabrication industry; continuing to intensify infrastructure construction; and making a quick and total development of modern service industry.

The "software" construction is intensified and the level of administration and service of the industrial park is improved, while quickening the "hardware" construction of the industrial park. For example, the dynamic administration information system for the industrial park development is established to implement the dynamic administration of the industrial park construction and layout with "3S" technology, and the agency service and pioneering service organizations such as market investigation, technical consultation, technical outcome transfer trade center, logistic center, etc are standardized and developed.

References:

[1] Yang Yuanfen, Analysis of Industrial Park Region Choice [J], Development Forum, 2005 (187): 63

[2] Shi Yishao, Li Shuangyan, Issues and Countermeasures on Land Utilization in Industrial Parks in Delta of Changjiang River [J], Compiled City Planning Publication, 2004 (4): 27-31

[3] Lu Xiaobo, Strategic Study of Planning Layout of Industrial Parks in Dalian [D], Dalian: Liaoning Teacher University, 2005.

[4] Teng Zhen, Deployment of Ecological Industrial Park Construction in Dalian Development Zone, Director Focus, 2002 (5): 26.

Studies on Technological Efficiency of Chinese Ecological Ports based on Neural Network Approach

Li Hezhong and Chen Shuwen

(School of Management, Dalian University of Technology, Liaoning, 116024)

Abstract: According to the sustainable development of Chinese ecological ports, a set of index evaluating the technological efficiency of ecological ports is forwarded in the beginning, to compensate the deficiency in index selection and the lack of statistical and quantitative studies in current related researches. In the following an appraisal model of ecological port technological efficiency is formulated based on neural network approach, to facilitate the evaluation of ecological port technological efficiency and the identification of major affecting factors. Empirical study is conducted on eight principal ecological ports in China including Dalian. Technological efficiency of each ecological port is calculated and scored. Affecting factors are identified and analyzed. Suggestions on how to improve the technological efficiency in Chinese ecological ports are given in the end.

Keywords: Ecological port, Technological Efficiency, Neural Network, Appraisal Model

1 Introduction

As the sustainable development of Chinese ecological ports goes on, the ecological ports' technological efficiency becomes more important in ecological ports operation. The ecological ports' technological efficiency (TE) refers to the ability to obtain optimal output with minimum input of resources into the ecological port. It is a measurement of the efficiencies of ecological ports in minimizing the input or in maximizing the output. With Chinese entry into WTO, the rapid development of ecological ports operations and the strengthening of services diversification, the competition among ecological ports are getting more and more stringent and competition of ecological ports efficiency will become the core of future competitions. At the same time,

Authors Introduction: Li Hezhong, Male, Ph.D candidate of Management school of Dalian University of Techonology - Chen Shuwen, Male, Born in March 1955 - Doctoral supervisor of Management school of Dalian University of Techonology.

technological efficiency plays a more and more important strategic role in various efficiency competitions of ecological ports [1].

There are two major methodologies in analyzing ecological port TE in current researches home and abroad. One is non-parameter methods and the other is parameter methods.

Data Envelopment Analysis (DEA) is the major representative approach in non-parameter methods [2,3,4]. Jose Tongzon (2001) [2] adopted a DEA model to assess the ecological ports' TE in Australia. Chen Junfei et (2004)[3] took the amount of circulating stocks as the input indexes, and benefits per stocks as the output indexes, and applied a DEA model to assess the relative operational efficiency of fifteen listed ecological ports transportation companies Yang Hualong etc (2005) improved a DEA model by using the berth length of container ports as input/output index[es] and made an empirical study on the efficiency of container ports in China. The non-parameter DEA approach can be used to assess indexes with different dimensions. There is no need for assigning the relative weightings for each indicator by the researchers so as to ensure a high objectiveness. But the minimum samples size must be twice that of the indexes in DEA approach [5].

Parameter methods mainly include linear regression methods [6], such as Stochastic Frontier Analysis (SFA) [7,8]. Jose L. Tongzon (2001) used the cost of ecological ports as the independent variable and the bridge suspending efficiency as dependant variable, and studied the synthesis efficiency of ports through regression analysis. Cullinane and Song (2002) [7] adopted SFA model in appraisal the efficiency of major container ports in Asia. Guo Hui (2005) [8] assessed the efficiency of some container docks of China and some other countries through Bayes SFA model, reaching such conclusions that the bigger the scale is, the greater the TE is, and that Chinese container ports efficiency as a whole is (lower) compared with that of the other countries in the world. Parameter methods are taking the errors caused by such factors as statistical, observational, and other ones into consideration. But the frontier functions assumed to have a certain subjective character [9], resulting in inaccuracy of the final appraisal results.

Despite big progress attained by current research, there are still some problems as shown in the following: (1) lack of study on appraisal index of Chinese ecological port TE, causing the

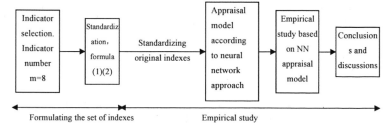

Fig.1 Research Framework

disability in reasonably reflecting the ecological port TE in China. For example, cargo transportation has a big proportion in total throughput of Chinese ecological ports at present. But almost all the current studies on ecological port TE ignore the ability of cargo throughputs [2]. The appraisal methods in current researches are influenced seriously by researchers' subjective factors and blurry stochastic factors. The values of ecological port TE are calculated in current studies on TE home and abroad. But no quantitative analysis has been made on major affecting factors of ecological port TE.

Considering all the above problems as a whole, this paper forwards a set of indices for appraisal of ecological port TE, basing the evaluation of ecological port output capacity on the total throughput ability of ecological ports in China.

2 Research Framework

According to indexes and data comprehensively collected in terms of ecological port TE assessment, a set of appraisal indices of ecological port TE is forwarded. After this, an appraisal model is set up based on neural network approach, and an empirical study on ecological port TE is conducted accordingly. This appraisal model could contribute to improve the scientific accuracy of the assessment, and to reduce the influence of subjective factors and uncertain stochastic factors as well. The empirical study proves this approach to be an intelligent method of appraisal ecological port TE. The final appraisal results are satisfactory. Pertinent suggestions are given on how to improve ecological port TE in the end. The whole framework is shown in fig. 1.

3 Indexes Selection

3.1 Existing Index

All the output and input indexes for appraisal of ecological port TE from various aspects mentioned in literature [2] [3] [4] [6] [7] [8] and [10] are summarized as shown in table 1.

There is already big progress attained in current researches. But here is still a dispute in defining indexes of ecological port output and input. Literature [3] starts the study from the financial point of view so that the operational efficiency of the listed companies could be embodied very well, but the final results would not be accurate enough if there are several companies who are operating in the same one port. Literatures [2] [4] [6] [7] and [8] include some indexes which are suitable for international ecological ports but not fit to China, lacking such important indexes as berth length and amount of cargo throughput etc. This insufficiency would lead to an error in the appraisal results and are not supportive enough to give pertinent suggestions on ecological ports development in China.

Tab.1 Existing Indexes of Ecological port Appraisal

Literature	Indexes
[3]	**Input:** (1)Asset(2) numbers of stocks in the market(3)numbers of workers(4)costs of core business **Output:** (1)benefits per stock(2)net interests(3)income of core business
[10]	(1)financial indexes(2)internal process(3)internal operations(4)indexes for learning and innovation
[2][4][6][7][8]	**Input:** (1)berth length(2)dock area(3)dock CFS(4)bridge suspending amount (5)postponing period **Output:** (1)containers throughput (2)growing rate of containers

3.2 Indexes selection

Considering the current situation of Chinese ecological ports and the ecological port synthetic throughput capacity, a set of ecological port TE appraisal indexes is forwarded accordingly. Four input indexes are mainly the berth length, dock area, bridge suspending amount and dock depth. Considering that in China cargo transportation has a big proportion in the whole throughput of ecological ports at present, output indicators are manly including the containers throughput, annual growth rate of port containers, port cargo throughput and annual growth rate of cargo throughput. Table 2 is the appraisal indexes of ecological port TE of China.

Tab. 2 Ecological Port Technology Efficiency Appraisal Indexes

Ecological port	Indexes
Input	Berth length
	Dock area
	Ecological port bridge suspending
	Port depth
Output	Containers throughput
	Annual growth rate of containers throughput
	Cargo throughput
	Annual growth rate of cargo throughput

The indexes selected in this paper are principally based on various fundamental inputs and throughput outputs of ecological ports and scientifically include those indexes such as port depth and cargo throughputs, which have important impact on Chinese ecological port TE into our index list. Firstly, this could compensate the deficiency of indicator selection and the lack of statistical and quantitative analysis in current researches. Secondly, this solves problems of incompletion in former studies due to using one single index only. Thirdly, this work could be a basis for a more in depth scientific appraisal of Chinese ecological port TE.

4 Appraisal model based on Neural Network

4.1 Theories of Neural Network

Neural Network (NN) is a kind of intellectual model[11]. The advantages of distributional information store, parallel treatment and self-organizing ability are contributing broadly to its application in the field of model identification, intellectual control, systematic modeling. Backpropagation (BP) is a representative approach of NN. The evolution of BP NN is based on the approach of reverse distributing arithmetic, which is a multilevel feedback network with the reverse error distributing and is composed of nodes of the input level, connotative level and output level. The basic idea of this arithmetic is to adjust the weighting and threshold values reversely according to the error of output level until the average square error goes to the minimum. After completing one "training process" using a set of training samples, the weighting values gained by the network would be the correct internal expression according to self-adapting training. Then we input the characteristic data of the experiment sample to the trained NN, the sample attributes will be justified and identified automatically by the NN. The structural principle is shown in fig. 2..

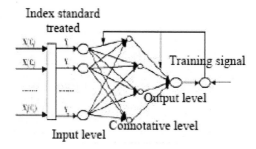

Figure 2 the structure of NN

4.2 Model setting

Based on the indexes selected above, an assessment model of ecological port TE will be set up according to BP NN approach as the following steps:

(1) Standardizing indexes dimensions

Dimensions of the selected indexes are different and need to be standardized. Values of normal indexes are treated in the following way:

$$F_j = (x_j - x_{j\min})/(x_{j\max} - x_{j\min}) \tag{1}$$

Values of reverse indexes are treated as:

$$F_j = (x_{j\max} - x_j)/(x_{j\max} - x_{j\min}) \tag{2}$$

F_j is the average of x_j, x_{jmin} is the minimum value of the pre-selected index j, x_{jmax} is maximum value of the pre-selected index j, j is the number of indexes.

(2) Design of assessment model structure based on NN

Input level: According to the assessing indexes system of ecological port TE we take the amount of indexes in the bottom level as that of nerve center of input level, which is nine in this study. Connotative level: The amount of neutral centers of connotative level has an important affect on the accuracy and learning efficiency of the whole BP NN. We take it as five through combining theoretical analysis and experiences. Output level: The appraisal of ecological port TE is a process from quantitative to quantitative. Appraisal result of ecological port TE is the output of the BP NN model. So we take the amount of neutral center of output level as one. In general, a 9-5-1 BP NN model is set up.

(3) Learning of BP network

The standard value of the ecological port sample training should be decided on first. The standardized samples of ecological port TE indexes are input into the BP NN. Then the learning process starts according to BP arithmetic. And the relative weighting of the neutral centers in each level and also the final training results would be gained.

(4) Appraisal of ecological port TE

To input the collected ecological port data into the BP NN to get the weighting outputs according to the learning step (3). Appraisal value of ecological port TE would be got as a consequence.

The ability of self-learning of NN is efficient enough so as to avoid the affection caused by subjectively stochastic factors. Besides, the ability of self-organizing of NN is also very strong to make it possible to escape the affection of fuzzy stochastic factors. Thus the side-effect of both above factors could be reduced by using NN approach to appraise ecological port TE.

Tab. 3 Original Data of China Ecological ports Technology Efficiency Appraisal Indexes

Year	Ecological port	Input indexes				Output indexes			
		Berth Length (meter)	Dock square (square meter)	Bridge suspending amount (piece)	Port depth (meter)	Container throughput (10,000 TEU)	Growth rate of container throughput	Cargo throughput (10,000 ton)	Growth rate of cargo throughput
2005	Dalian	2335	102.6	12	9.0	151	1.29	9446	1.22
	Guangzhou	2934	89.2	10	8.0	238.36	1.407	12288	1.106
	Ningbo	1984	70.4	6	18.0	276.79	1.286	15819	1.201
	Qingdao	2879	210.5	24	10.0	357.29	1.24	10764	1.139
	Shanghai	4764	158.3	46	10.0	1015.73	1.268	24305	1.112
	Shenzhen	3490	140.9	31	15.5	881.02	1.222	8645	1.158
	Tianjin	2360	113.0	20	9.0	268.74	1.248	13907	1.209
	Xiamen	3126	163.4	22	10.0	190.38	1.158	2677	1.123
2006	Dalian	2335	102.6	12	9.1	175.85	1.164	10427	1.104
	Guangzhou	2934	89.2	10	8.0	360.67	1.504	17195	1.393
	Ningbo	2138	75.7	8	18.0	383.33	1.370	23581	1.122
	Qingdao	3032	220.5	28	10.1	427.58	1.197	12843	1.193
	Shanghai	4764	163.5	46	10.0	1198.21	1.179	26877	1.105
	Shenzhen	3700	169.4	34	15.5	992.91	1.125	9569	1.108
	Tianjin	2360	113.0	20	10.0	329.95	1.228	15044	1.082
	Xiamen	3126	163.4	23	10.0	219.83	1.155	4239	1.571

Data sources: 1. Data on Container throughput, Growth rate of container throughput, Cargo throughput, Growth rate of cargo throughput is according to the website of China Transportation. See :http://www.moc.gov.cn/;

2. Other data such as berth length are referred to the latest literature [4] and China port association website: http://www.port.org.cn.

5 Empirical Study

5.1 Data

Empirical study is made on 8 main ecological ports including Dalian, Qingdao, Tianjin, Shanghai, Ningbo, Xiamen and Shenzhen. Data of these ecological ports in July 2005 and July 2006 are shown in table 3.

5.2 Appraisal of ecological port TE

5.2.1 Standardization of indexes dimensions

To meet the needs of NN, dimensional standardization of the original data in table 3 is carried out first. As shown in table 2 and 3, in the indexes system, both the input and the output indexes are the normal ones. So the standardization is made according to formula (1). The results are shown in table 4.

Tab. 4 Standard Data of China Ecological ports Technology Efficiency Appraisal Indexes

Year	Port	Input indexes				Output indexes			
		Berth Length	Dock square	Berth Length	Dock square	Container throughput	Growth rate of container throughput	Cargo throughput	Growth rate of cargo throughput
2005	Dalian	0.1263	0.2298	0.1500	0.1000	0.0000	0.5301	0.3130	1.0000
	Guangzhou	0.3417	0.1342	0.1000	0.0000	0.1010	1.0000	0.4444	0.0000
	Ningbo	0.0000	0.0000	0.0000	1.0000	0.1455	0.5141	0.6076	0.8333
	Qingdao	0.3219	1.0000	0.4500	0.2000	0.2386	0.3293	0.3739	0.2895
	Shanghai	1.0000	0.6274	1.0000	0.2000	1.0000	0.4418	1.0000	0.0526
	Shenzhen	0.5417	0.5032	0.6250	0.7500	0.8442	0.2570	0.2759	0.4561
	Tianjin	0.1353	0.3041	0.3500	0.1000	0.1362	0.3614	0.5192	0.9035
	Xiamen	0.4108	0.6638	0.4000	0.2000	0.0455	0.0000	0.0000	0.1491
2006	Dalian	0.0750	0.1858	0.1053	0.1100	0.0000	0.1029	0.2733	0.0450
	Guangzhou	0.3031	0.0932	0.0526	0.0000	0.1808	1.0000	0.5723	0.6360
	Ningbo	0.0000	0.0000	0.0000	1.0000	0.2029	0.6464	0.8544	0.0818
	Qingdao	0.3404	1.0000	0.5263	0.2100	0.2462	0.1900	0.3801	0.2270
	Shanghai	1.0000	0.6064	1.0000	0.2000	1.0000	0.1425	1.0000	0.0470
	Shenzhen	0.5948	0.6471	0.6842	0.7500	0.7992	0.0000	0.2354	0.0532
	Tianjin	0.0845	0.2576	0.3158	0.2000	0.1507	0.2718	0.4773	0.0000
	Xiamen	0.3762	0.6057	0.3947	0.2000	0.0430	0.0792	0.0000	1.0000

The training standard value of 2005 is set according to current studies on ecological port TE in literature [3] [4] and [8]. Value 1 is in the first group standing for excellent ecological port TE. Value 2 is in the second group standing for normal ecological port TE. Value 3 is in the third group standing for worse ecological port TE. Please refer to the second column in table 5.

Tab.5 2005 Year Training Standard data and Results

Ecological port	Training standard Value	Training value of 2005
Dalian	-1	-0.9974300
Guangzhou	1	0.9987000
Ningbo	1	0.9989200
Qingdao	0	-0.0020017
Shanghai	1	0.9997100
Shenzhen	-1	-0.9990000
Tianjin	0	-0.0043749
Xiamen	0	-0.0038273

5.2.2 NN learning

Then the data of 2005 on ecological port TE appraisal listed in table 3, as the sample data, are input into the Matlab software of BP NN to finish the learning process. Training values of 2005 are gained as shown in the third column of table 5.

5.2.3 Appraisal of ecological port TE

Then we input the standardized data on ecological port TE appraisal of 2006 listed in table 4 into the BP NN. The results are calculated based on the sample training values of 2005 listed in table 5 See table 6.

Ratings of Chinese ecological port TE are: Ningbo port, Shanghai port, Tianjin port, Dalian port, Qingdao port, Guangzhou port, Shenzhen port and Xiamen port, shown in table 6.

Ningbo port rates the No. 1 due to the high level of its outputs and also its inputs. Shanghai port advantages are apparently greater of its berth length, bridge suspending and throughput amounts, resulting higher input and output values. Both the scale and throughput amount of Xiamen port are lower than the other ecological ports, leading to the lower input and output values and consequently the last rating of Xiamen. Although the input value of Shenzhen port is bigger, its output is smaller, resulting relatively lower integrated output/input capabilities. So Shenzhen port gets to be the last second.

Generally speaking, Chinese ecological ports should look closely at their own situations, identify the crucial factors affecting, and solve the key problems to get ecological port TE improved.

Tab.6 Technology Efficiency Evaluation of China Ecological ports

Ecological port	Appraisal results	Rating
Dalian	0.62718	4
Guangzhou	-0.97262	6
Ningbo	0.99963	1
Qingdao	0.39446	5
Shanghai	0.99945	2
Shenzhen	-0.86489	7
Tianjin	0.90793	3
Xiamen	-0.99992	8

6 Conclusions

(1) Closely footing on the current situations of Chinese ecological ports and their technological abilities, an appraisal indexes system of ecological port TE is set up according to NN approach, to compensate the deficiency in indexes selection and statistical and quantitative methods in existing related studies.

The appraisal indexes system of ecological port TE has taken various basic input and throughput output of ecological ports into consideration and scientifically include those important affecting factors such as ecological port depth and cargo throughputs into the appraising system. This has contributed to compensate the deficiency in indexes selection and statistical and quantitative methods in existing related studies on the one hand and to solve the problems of incomplete appraisal and analysis on the other hand.

(2) Appraisal model based on NN approach is set up to assess objectively the ecological port TE. Empirical study is made on eight main ecological ports in China including Dalian port to calculate the values of each ecological port TE and to rate them accordingly.

The results of empirical study prove that Ningbo port and Shanghai port is the No. 1 and 2 respectively. Shenzhen port rates at the last second due to its lower input/output values. Xiamen port rates at the last one because both its scale and throughputs amount are lower than all the other ports. This study would help Chinese ecological ports identify crucial factors affecting their TE and to take pertinent measures to get TE improved.

(3) Pertinent suggestions are given on how to improve ecological port TE in China.

Chinese ecological ports should improve the infrastructure qualities. They should increase the ability of ecological port cargo throughputs and at the same time greatly enhance that of containers throughputs to higher their outputting efficiencies. Meanwhile ecological port multi-services should also be highly attached importance to increase ecological ports outputting efficiencies.

References

[1] WANG TengFei, Cullinane, K., etc. Container Port Technology and Economic Efficiency [M], *Palgrave-Macmillan*, 2005.

[2] Jose Tongzon. Efficiency measurement of selected Australian and other international ports using data envelopment analysis [J], *Transportation Research Part A*, 2001, (35): 113-128.

[3] Chen Junfei, Xu Changxin, Yan Yixin. Appraisal fo operating efficiency of listed companies in port water transportation based on data envelopment analysis[J].Journal of Shanghai Maritime University, 2004, 25, 1, 51-55.

[4] Yang Hualong, Ren Chao, Wang Qingbin. Appraisal of container port performance based on data envelopment analysis[J]. Journal of Dlian Maritime University, 2005, 31, 1, 51-54.

[5] Shen Shaohan, Zhu Qiao, Wu Guangmo. DEA Theories, methods and applications [M]. Scientific Publication. 1996, 22-41, 65-72,155.

[6] Jose L. Tongzon. Determinants of port performance and efficiency [J]. *Transportation Research Part A*, 2001, 29A, (3): 245-252.

[7] Kevin Cullinane, Dong-Wook Song, Richard Gray. A stochastic frontier model of the efficiency of major container terminals in Asia: appraisal the influence of administrative and ownership structures. *Transportation Research Part A* 2002, (36) :743–762.

[8] Guo Hui. Analysis on TE of container port – an comparing of Chinese container ports with their counterparts in the world. Master thesis of Dalian Maritime University, 2005. 3.

[9] Kevin Cullinane, Teng-Fei Wang, etc. The technical efficiency of container ports: Comparing data envelopment analysis and stochastic frontier analysis [J]. *Transportation Research Part A*, 2006, (40):354–374.

[10] Long Ruizhi, Li Guping, Analysis and discussion on port performances appraisal[J]. Port Economy, 2005.5, 43-44.

[11] Bruck J. A study on neural networks [J].International Journal of Intelligence Systems, 1988, (3):122-149,148.

Sustainable Livelihood Approach

in connection with the construction of

Eco-industrial Park in China

Yuan Bin[1], Yu Chengxue[2]

(School of Management, Dalian University of Technology, Liaoning Province, China, 116024)

Abstract: The construction of eco-industrial Park needs more land, which makes many farmers lose their land, and their livelihoods being influenced seriously. This paper firstly introduces the Sustainable Livelihoods Approach (SLA), and then analyses the disadvantages of the current compensation for Chinese landless farmers. At last it promotes a new compensation mode for the landless farmers for their sustainable livelihoods.

Keywords: Eco-industrial Park; Land lost farmers; Sustainable livelihoods

1 Introduction

Without repeating various theoretical views of sustainable development and the related areas of industrial ecology, design manufacturing or symbiosis, sustainability in the context of businesses now has an increasingly strong backing and support from both public policy and private sectors [1]. This is a significant issue debated by most corporate Boards of Directors under one form or another of "corporate governance". That is economic statistics and decisions. And what does it mean for a society to be sustainable or sustained? What should the society attempt to sustain? Articles on such questions impress, not by progress toward the goal of sustaining society, but by the vagueness of the concept. In a frequently cited statement, the Brundtland (1987) report argues for a balance between the needs of the present and of the future. But those needs are not specified and, more importantly, the balance to be implemented is not defined [2].

In the intervening years, the notion of sustainability has come to encompass all that the proponent thinks would be ideal in a society, including environmental protection, climatic stability,

---

[1]   Yuan Bin. PhD Candidate, School of Management, Dalian University of Technology, Liaoning Province, China, 116024

[2]   Yu Chengxue. PhD, School of Management, Dalian University of Technology, Liaoning Province, China, 116024; E-mail: chengxueyu@163.com

elimination of poverty, concern for the well-being of future generations, good corporate and social governance, social inclusiveness, maintenance of small communities, ways of life, and others.

In the current society, we need the sustainability. Sustaining means many things to many people. Any definition of sustainability, however, must encompass what is known as in the generation equity. In analyzing the use of an exhaustible resource, oil, through time, Solow (1974) applied the following criterion: A society's success should be judged by the standard of living of the least fortunate generation looking forward from the present. His objective, called the maximum objective, was to maximize that minimum standard of living, an objective that contrasted starkly with the utilitarian one of maximizing present value [3].The world needs sustainable development. And up to now, the problems of environment often happen in the developing countries. The goal of developing countries is an extension of sustaining that encompasses increasing consumption until a "developed" state is reached. Mink(1993) discusses links between poverty and environmental exigencies, and policies to break out of the trap of poverty with a degraded environment [4]. Solow (1993) is clearly content [5].

Since the 1970s, natural resources and the environment have been the main motivators of extending the national accounts, especially to extend NNP to incorporate all measurable contributors to social well-being. They have served a similar purpose in discussions of sustaining the economy or of sustainable development. 1994 marked the twenty-fifth anniversary of the Apollo 11 lunar mission, which provided the first opportunity for humans to view the Earth from the Moon. Thanks to television, millions of people had the physical experience of seeing our home, the Earth, as a whole island of life in the vastness of space. Among our fundamental learning from this vision is that there is no 'away' in which to throw our wastes and that our home's resources are indeed finite. These insights have led to diverse and contradictory attempts to define sustainable development [6]. Central to only some definitions of this concept is an idea derived from the field of ecology-carrying capacity. Ecologists define this as the population of a given species that can be supported indefinitely in a defined habitat without permanently damaging the ecosystem upon which it is dependent. William Rees defines human carrying capacity as 'the maximum rate of resource consumption and waste discharge that can be sustained indefinitely without progressively impairing the functional integrity and productivity of relevant ecosystems (wherever the latter may be)'. Industrial ecology suggests a means of linking these valuable innovations into a larger system, the interaction of industrial systems with the biosphere. The embedded concept of industrial ecosystems integrates cleaner production into the interactions of companies in a specific industrial region or park with its local and ultimately global ecosystem. Industrial ecology may be defined briefly as the means by which our species can deliberately and rationally approach and maintain a desirable global carrying capacity. Industrial ecology suggests using the design of ecosystems to guide the redesign of industrial systems. The goal is to achieve a better match of industrial performance with ecological constraints.

The ultimate reason for sustainable development is the serious pollutions caused by industries, which need to be changed to eco-industrial Park for symbiosis. The construction of eco-industrial Parks needs more land firstly, and then needs many linking firms gathering together for symbiosis. For this reason, all the residents there have to move away, thus, the farmers there will lose their land they depend on for living. So, the local government should offer certain money and other advantage policies to them for compensation. And now, many eco-industrial parks have been established in China, so, Chinese government has to consider the profits of the land lost farmers. And the current way cannot meet the demand of the farmers, they still cannot live. This paper tries to give some measures to solve the problems as follows.

Firstly, this paper explains the sustainable livelihoods; and secondly, it states the disadvantages of the current Chinese compensation for land lost farmers, and then, it implies the framework of sustainable livelihoods to solve the problems of the Chinese land lost farmers; and thirdly, it brings out a new mode of sustainable livelihoods for Chinese land lost farmers. Lastly, it gives a conclusion to conclude the whole paper.

2 Methodologies

2.1 Sustainable Livelihoods Approach (SLA)

The Sustainable Livelihoods Approach (SLA) is prominent in recent development programs that aim to reduce poverty and vulnerability in communities engaged in small scale fishing, fish processing and trading [7,8]. The 'capitals and capabilities' framework [9,10] underpinning the SLA has been used to design research into the role of small-scale fishing in rural economies and to inform policy that seeks to enhance the contribution of small-scale fisheries to poverty reduction and improve livelihoods and security of fisheries-dependent people [11,12,13]. The SLA is, however, an approach to development policy and practice, not just a research or conceptual framework. The definition and measurement of poverty has evolved from a focus on low income and consumption, first to encompass a lack of basic needs (access to food, shelter, health and sanitation), then to include a lack of basic human rights, and finally to reflect more qualitative understandings that capture peoples' own experiences and definitions, including psychological aspects such as feelings of powerlessness, humiliation and insecurity [14]. The literature on social exclusion [15] adds a political dimension by focusing on how, through their relations to the more powerful, groups of people may become excluded from economic opportunities, social networks and political processes. Related to these definitions is the concept of vulnerability (of a person or livelihood), which is a function of the risks to which people may be exposed, the sensitivity of their livelihood system to those risks, and their ability to adapt to, cope with, or recover from the impacts of an external 'shock' to their livelihood system [16]. Understanding and responding to these multiple dimensions of poverty requires a broad, multi-disciplinary approach. The SLA is useful in this context as it provides both a set of guiding principles and an analytical framework.

2.2 The SLA Principles

The core principles that underlie SL thinking can be summarized as follows [17]:

1. Putting people's social and economic activities at the centre of the analysis;
2. Assessing options for management and development intervention that transcend sectoral boundaries;
3. Making micro-macro links;
4. Being responsive and participatory in a addressing management priorities;
5. Building on strengths;
6. Taking a broad view of sustainability.

While none of these principles are new or unique to the livelihoods approach, taken together, they represent a new way of working in development that have yielded positive results in other areas of rural and natural resource development [18, 19].

2.3 The Livelihoods Framework

The livelihoods framework brings together assets and activities and illustrates the interactions between them (Fig. 1). The social and economic unit considered is typically represented by the household, conceived as the social group which resides in the same place, shares the same meals and makes joint or coordinated decisions over resource allocation and income pooling.

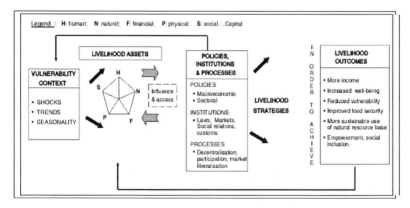

Fig. 1. The rural livelihoods framework as a means to understand natural resource management systems.
(Source: modified from UK Department for International Development.)

The capital assets owned, controlled, claimed, or by some other means accessed by the household are grouped into five categories. These comprise physical capital (at household level-boats, house, bicycle etc. but also at community or citizen level, access to infrastructure such as harbors, road networks, clinics, schools etc.); financial capital (savings, credit, insurance); natural capital (fish stocks, areas of seabed leased or accessed by license, land owned, crops cultivated etc.); human capital (people's 'capabilities' in terms of their health, labor, education, knowledge, skills and health); and social capital (the kinship networks, associations, membership organizations and peer-group networks that people can use in difficulties or turn to in order to gain advantage). Some argue that this framework would benefit from the addition of additional categories of capital-political and cultural [20].

Access to both assets and activities is enabled or hindered by policies, institutions and processes (PIPs), including social relations, markets and organizations. PIPs include access and rights regimes and how they work-or do not. Livelihood sustainability is also affected by external factors, referred to as the vulnerability context, comprising cycles (e.g. seasonality), trends and shocks that are beyond the household's control. Understanding how people succeed or fail in sustaining their livelihoods in the face of shocks, trends and seasonality can help to design policies and interventions to assist peoples' existing coping and adaptive strategies. These may include improving access to education and health care facilities, strengthening rights to land for settlement and agriculture, reforming local tax and license systems, providing financial and enterprise development services and promotion of diversification. Capital assets permit livelihood strategies to be constructed by individuals or households. Some authors object to the term 'strategy' for what they see as the outcome of a bundle of reactive and unplanned actions [21]. Finally, this Framework paints to outcomes. A livelihood is sustainable if people are able to maintain or improve their standard of living related to well-being and income or other human development goals, reduce their vulnerability to external shocks and trends and ensure their activities are compatible with maintaining the natural resource.

2.4 The status livelihoods of land lost farmers in China

With the rapid development of eco-industrial parks in China, more and more farmers will lose their land they depend on for living. According to the statistics, 40-50 million farmers have lost their land entirely or partially, and it will enhance to 20 millions per year [22].

The problem of sustainable livelihoods is fundamental. Therefore, the sustainable livelihoods can be regarded as a basic target to purchase the land. Here it means that a person or a family has the possibility to make a living, to own/have assets and some income they possess or gain.

Sustainable Livelihoods Framework (SLF) gives an integrated analyzing method to reduce poverty, that is to say the farmers should collocate their livelihood capital actively to maintain their current living. The current compensation mode for land lost farmers follows the item 47 of the land law of the People's Republic of China (Fig. 2).

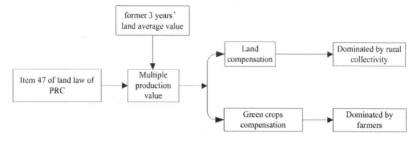

Fig. 2 the Current compensation mode for land lost farmers of China

This mode shows that it doesn't compensate to the farmers according to the real land price, but takes the criteria of many years production value before they purchase the land, it doesn't include the land value added, and it also doesn't consider the balance of different places, which makes the compensation lower, and makes the farmer loose their land, and at the same time loose their employment, especially for the old farmers. This influences the farmers' sustainable livelihoods.

2.5 Solutions for the land lost farmers

In order to solve the problem of the land lost farmers, we try to imply SLF to the land lost farmer during the construction of eco-industrial Park, and establish a new compensation mode for Chinese sustainable livelihoods (Fig. 3).

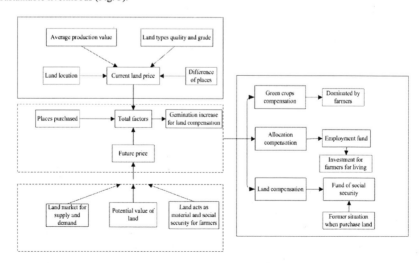

Fig. 3 the ideal compensation mode of sustainable livelihoods for Chinese land lost farmers

This new compensation mode tries to promote or renew the farmers' income and living level, it depends on the current marketing land price to compensate the farmers when they purchase the land for the construction of eco-industrial Parks. It also considers the land types, quality, grade, different places, potential value, land market for supply and demand, land acts as materials and social security for farmers and so on. This mode overcomes the disadvantages of the current compensation mode.

3 Conclusion

This paper tries to introduce the eco-industry and symbiosis at the beginning, then it introduces the construction of eco-industrial parks, because the eco-industrial park needs more land, which lead to a situation where many farmers lost their land. Even though the Chinese government gives certain compensation to them their livelihood is influenced seriously. This paper tries to overcome the disadvantages of the current compensation mode, and on this base it proposes a new compensation mode for land lost farmers.

Reference:
[1] Woodrow Clark, Henrik lund. Sustainable development in practice. Journal of Cleaner Production 15 (2007) 253-258.
[2] Brundtland (World Commission 1987).Our Common Future, Oxford University Press, Oxford, New York.
[3] Solow, R. M., 1974. Intergenerational Equity and Exhaustible Resources. Review of Economic Studies, Symposium Issue.
[4] Mink, S., 1993. Poverty and environment. Financ. Dev., 8-9.
[5] Solow, R. M., 1993. An Almost Practical Step Toward Sustainability, Resources for the Future, Washington DC, reprinted in Resources Policy 19, 3, September 1993, 162-172.
[6] Ernest A. Lowe and Laurence K. Evans, 1995. Industrial ecology and industrial ecosystems. Journal of Cleaner Production. Vol. 3, No. 1-2, 47-53.
[7] Neiland AE, Bènè C, editors. 2004.Poverty and small scale fisheries in West Africa.Dordrecht, Netherlands: Kluwer, Rome, Italy: FAQ; 2004.
[8] Stirrat RL. Yet another `magic bullet':the case of social capital. Aquatic Resources, Culture and Development 2004:1(1):25-33.
[9] Chambers R, Conway GR. Sustainable rural livelihoods: practical concepts for the 21st century. Institute of development studies discussion paper no. 296. Brighton, Sussex, UK; 1992.
[10] Bebbington A.Capitals and capabilities: a framework for analysing peasant viability, rural livelihoods and poverty. World Development 1999;27(12): 2021-44.
[11] Allison EH, Ellis F. The livelihoods approach and management of small-scale fisheries. Marine Policy 2001;25(5):377-88.
[12] Allison EH.The fisheries sector,livelihoods and poverty reduction in eastern and southern Africa. In: Ellis F, Freeman HA, editors. Rural livelihoods and poverty reduction policies. London: Routledge; 2005. p.256-73.
[13] Bènè C, Macfadayen G, Allison EH.Enhancing the contribution of small-scale fisheries to poverty alleviation and food security. Fisheries Technical Report No. 481, FAO, Rome; 2006.
[14] Narayan D, Chambers R, Shah MK, Petersch P. Voices of the poor: crying out for change. Oxford:Oxford University Press:2001.
[15] Sen A, Social exclusion: concept, application, and scrutiny. Social development papers no. 1. Manila, Philippines: Office of Enviroment and Social Development, Asian Development Bank; June 2000.

[16] Adger WN, Brooks N, Bentham G, Agnew M, Eriksen s.New indicators of vulnerability and adaptive capacity.Technical Paper No.7, Tyndall Centre for Climate Change Research, University of East Anglia, Norwich, UK; 2004.128pp.

[17] Edward H.Allison, Benoit Horemans. Putting the principles of the Sustainable Livelihoods Approach into fisheries development policy and practice. Marine Policy 30 (2006) 757-766.

[18] Ashley C, Carney D. Sustainable Livelihood: Lessons from Early Experience. London: Department for International Development; 1999.

[19] Neely C, Sutherland K, Johnson J. Do sustainable livelihood approaches have a positive impact on the rural poor? A look at twelve case studies. Livelihood support programme, Working paper 16. Rome: Food and Agriculture Organisation; 2005.

[20] Stirrat RL. Yet another 'magic bullet': the case of social capital. Aquatic Resources, Culture and Development 2004; 1(1): 25-33.

[21] Dorwood A, Poole N, Morrison J, Kydd J, Urey I. Markets, institutions and technology: missing links in livelihoods analysis. Development Policy Review 2003; 21(3): 319-32

[22] Group of question for discussion of Chinese Social science academy. The sustainable livelihoods countermeasures for land lost farmers [N]. Economy Reference Newspaper, 2004,12,25.(in Chinese)

2.3

Examples and Case Studies

Science and Technology Park Management: From Business Integration to Technology and Urbanity - The Case of Berlin-Adlershof[1]

Harald A. Mieg and Ulrike Mackrodt

Georg-Simmel Center for Metropolitan Studies, Humboldt-Universität zu Berlin

Abstract: This chapter discusses Science and Technology Park Management (STPM). We argue that in the future the added value and competitive advantage provided by STPM will not only arise from successful business-integration management but, more importantly, from technology clustering on the one hand, and through the provision of an attractive urban environment ("urbanity management") on the other. This article introduces the Science and Technology Park (STP) Berlin-Adlershof. Discussion centers on the change of the type of clusters represented by STPs and the professionalization of STPM.

Keywords: Science and Technology Park Management, Berlin-Adlershof, urban environment, technology clusters, professionalization

1. Science and Technology Park Management

Science and technology parks (STPs) are interfaces between industry, universities and other research institutions. Usually, STPs are managed. In the following, we will refer to the general task or work of STP managers as science and technology park management (STPM). Coordination and the support of business incubation are the main instruments of STPM. In general, STP managers are translators and connectors, who translate research into business and vice versa.

There are several approaches to describe the functioning of STPs. The one cited most often is the triple-helix concept by Etzkowitz and Leydesdorff (2000). The triple-helix concept describes the development of STPs within a network of university-industry-government relations. In the following, we would like to emphasize the managerial component of STP development. Our two

---

[1] The preparation of this chapter was supported by the HANS-SAUER-STIFTUNG, a German inventors foundation.

basic assumptions are: Firstly, successful STPs need to be managed. Secondly, there is a growing global competition of STPs. Thus STPM needs to take this global perspective into account and aim to create and enhance strategic advantages of the particular local STP.

Cabral (1998) defined a managerial "paradigm" for STPM which based on empirical studies on science parks. This paradigm consists of a ten-point list that identifies crucial tasks for the successful development of STPs. We can divide between internal and external managerial tasks[2]:

Internal tasks of STPM are

- To provide marketing and management support to firms located in the STP
- To create and analyze the potential market for the STP's products and services
- To provide protection for product/process secrets through patents etc.
- To establish financial expertise to develop the STP economically and sustainably
- To select suitable firms to join the STP
- To integrate consulting firms within the STP

External tasks are

- To create a recognizable identity for the STP
- To obtain the support of a powerful and stable partner in the local or national economic/political environment
- To include a high profile person within the STP management board in order to give the park's interests a face and voice

General task (internal and external):

- To provide access to qualified personnel

The task list by Cabral underpins both the importance of university-industry-government relations as described by the triple-helix concept and the central role of the STP manager. In particular, the Cabral list is a product of the professionalization of the STPM business. This professionalization is reflected through the foundation of professional associations such as the EBN (European Business and Innovation Centre Network), the ASTP (Association of European Science & Technology Transfer Professionals) or the IASP (International Association of Science Parks). It goes along with the establishment of standards for STPM which are promoted and taught in conferences and professional training courses, for instance by ASTP. Knowledge on managerial aspects of STPM is exchanged through the presentation of cases and managerial experiences (cf. IASP, 2008). Small wonder that in a study by Sanz with STP managers of the IASP (2008), networking turns out to be a main task of STP managers, resulting in 85 % of Sanz' sample belonging to two or more networks.

---

[2] Cabral (1998) did not differentiate between internal and external tasks. This differentiation has been added by the authors.

The professionalization of STPM - mainly with regard to providing business incubation and a joint marketing/branding for a particular STP - has led to STPs competing at a homogenous high level with regard to the services and facilities provided. This development has been reinforced by the increasing number of STPs in Europe, Asia, and Northern America. Benchmarking STPs is about to become a professional STPM standard. The STP sector now witnesses a tendency towards equalization and exchangeability of STPs regarding the services provided by STPM. However, there is still a divide between the need for a global perspective and measures taken: 90 % of the managers interviewed by Sanz consider attracting foreign companies as a top priority but a majority of STPs spends less than 15 % of their marketing budget on international marketing (Sanz, 2008).

Global competition and professionalization will force local STPMs to focus on new fields and strategies in order to successfully compete for companies, people, and investment. We argue: The added value of a particular STP will no longer arise from successful business-integration management alone but from (1) technology clustering (joint technology development by enterprises and universities) and (2) "urbanity management," that is developing the attractiveness of the STP's urban environment. These two very different measures transcend traditional managerial tasks, integrating technological innovation on the one hand with urban development on the other.

1.1 Technology clustering

STPM aims at the support of the resident enterprises regarding organizational, financial or technological issues. From this point of view, the core task of STPM consists of network management, that is supporting the flow of knowledge, people, and investments.

The strategy of technology clustering goes one step further: Instead of only supporting the activities and development of enterprises and enterprise clusters, STPM actively engages in the process of technology advancement. Possible measures include holding conferences, organizing exchanges or inviting researchers and professionals from outside the particular STP. As uncertainty prevails in technology advancement and technological innovation, the overall task of a particular STPM should be to define and manage a portfolio of technologies. Thus the clustering of technological activities becomes a necessity.

The study by Sanz (2008) shows that STPs and STP managers remain uncertain about the role of technology clusters. About 40 % of the STPs define themselves as specialized on specific sectors, 30 % are more less generalists without a clear specialization. In sum, 55 % of all STPs still admit companies from all sectors.

1.2 Urbanity management

If a STP aims to become a driver in global technological development, this requires the attraction of highly qualified personnel (especially top scientists or engineers). These "high potentials" will have specific expectations regarding their job, salary as well as quality of life at their working site. Therefore, the core task of STPM is to attract these people as prospective employees and company founders.

As human resource management is the duty of the resident companies, STP managers can only help to improve the general contextual conditions of a particular STP. They do not directly influence salaries and bonuses but they can work on improving the working environment. Suitable areas of action include the provision of kindergartens and schools as well as leisure facilities such as access to parks. The ultimate goal would be the creation of an "urban" environment. The quality required is best described as "urbanity," that is the quality of living in an intact, well functioning neighborhood or city with diversified public and private services - at least some of them in walking distance.

The concept of an integrated campus that offers an urban image has already been realized in other contexts. Examples are the campus idea of American universities, the French *technopoles* or corporate park projects such as the Infineon Campeon in Munich, Germany. There are also some STPs with a strategy for urban management such as Recife, Brazil, (da Silva, 2008) or Wollongong, Australia, (Fuller, 2008). However, though STPs are an urban phenomenon with 86 % of the STPs located in cities (Sanz, 2008), to date urbanity management is not a common principle in the development of STPs.

2. The Berlin-Adlershof Case

The general considerations regarding STPM introduced in the first chapter are illustrated using the case of the STP Berlin-Adlershof, one of Europe's largest STPs. We will show, firstly, how the tasks and measures for a successful STPM formulated in Cabral's managerial paradigm have been realized. Secondly, referring to the recent history of Berlin-Adlershof, we explain how technology and urbanity shape the new STPM strategies of the Adlershof management. As there is hardly any literature on Adlershof (e.g., Herfert, 2005), we will start with some general remarks on the historical development of Berlin-Adlershof (for more detailed information see www.adlershof.de).

2.1 The historical roots of Berlin-Adlershof: Aviation and the GDR Academy of Sciences[3]

Already at the beginning of the 20th century, the area of the STP Adlershof was used as a site for science and technology development. The area where the STP Adlershof is located today was one of

---

[3] Parts of the description of the history of Adlershof are taken from www.adlershof.de.

the first German airbases. Germany's first motorized aircraft took off from there. In 1912, the German Experimental Institute for Aviation DVL (Deutsche Versuchsanstalt für Luftfahrt) made Adlershof its headquarter. Laboratories, motor test beds, wind tunnels and hangars were erected in the 1920s and 1930s and represent historical landmarks on the campus today.

After World War II, the airport was closed and Berlin-Adlershof situated in the south-east of Berlin, became part of the German Democratic Republic (GDR) during the German separation (1945-1990). The GDR, as a socialist country with a centralist economy, founded several large institutions in its capital, as well as in Berlin-Adlershof. Thus Adlershof became home to nine institutes of the GDR Academy of Sciences in the fields of physics and chemistry as well as to the GDR National Television. The Academy of Sciences represented an important research institution. A substantial portion of the research done at the Academy institutes was notable for its close collaboration with industry, allowing the scientists to engage in world-wide networks. Adlershof produced many known inventions, such as ultra-short pulse lasers, time-resolved optical spectroscopy and space diagnosis devices.

After the German reunification in 1990, the Academy of Sciences and all employees, more than 5,600, had to face the closure of their institutions. As similar research institutions already existed in West Germany, the former GDR institutes were dissolved. Every third employee was integrated with the existing research institutions of the Federal Republic of Germany but the rest were forced to either find new jobs in private enterprises or to start their own businesses.

2.2 The Science and Technology Park Berlin-Adlershof

2.2.1 The Project: Adlershof - an integrated city

The Science and Technology Park Berlin-Adlershof came into being in the early 1990s when an urban planning concept, created by the City of Berlin, defined the vision for the development of Adlershof. This concept did not only refer to the future STP but dealt with a larger area which also included the abandoned airfield and the former national television campus of the GDR.

In 1992, the Berlin Senate voted for the creation of an integrated location for science and industry in Adlershof. Since then more than 1.5 billion € have been invested in the location. About 230 million € were used for new construction projects. This urban development project was meant to integrate the tradition of science and media business in Adlershof. Therefore the Adlershof project has been called: City of Science, Technology, and Media.

The urban planning concept also included measures to increase the living quality of the area such as the transformation of parts of the former airfield into a landscape park and the construction of residential areas located along the edges of the landscape park. The aim was to develop an integrated city (BAAG, 2003), a concept that is also addressed as "the European city" (cf. Siebel, 2004). Figure 1 provides an overview of the plan and the current situation.

2.2.2 STPM

Today, the Berlin-Adlershof STP is home to around 400 high-tech enterprises – most of them small and medium sized – with more than 4,200 employees and eighteen scientific research institutes with over 2,300 employees.[4] Since its inception, Berlin-Adlershof has grown constantly (see Figure 2) and is now developing towards an integrated urban fabric.

The STP is managed by WISTA Management GmbH. WISTA Management is the developer and incubator and provides facility management for Berlin-Adlershof. Its main tasks are:

- Marketing of the Adlershof brand
- Attraction and selection of additional companies
- Definition of development strategies for Adlershof.

WISTA Management co-operates with the two Adlershof-based business incubation centers. One of them, the IGZ Adlershof (Innovations- und Gründerzentrum), has become Germany's biggest business incubation center. Moreover, WISTA Management runs the "Eurooffice" service, providing guest offices for visitors from foreign STPs and companies interested in locating to Adlershof.

The ongoing professionalization of Adlershof STPM is reflected for instance by international networking with other STPs. WISTA Management has established a co operation with STPs in the Baltic region (Baltmet Inno, www.inno.baltmet.org) as well as project co-operations with cities such as Amsterdam or Copenhagen. One of the involved tasks is to define benchmarks for STPM, another one is to discuss strategic objectives such as the development of creative industries.

2.3 Adlershof from the viewpoint of Cabral's managerial paradigm

In the following, we discuss the STP project Berlin-Adlershof from the point of view of three selected tasks from the Cabral list. We will see that the backbone of the STP Berlin-Adlershof are the people who work there and stick to the location, to their homebase: Adlershof.

2.3.1 Access to qualified personnel

In its first years of existence, the STP Berlin-Adlershof profited from the disposability of a great number of qualified personnel in the sectors of physics and chemistry due to the dissolution of the GDR Academy of Sciences. This proved a crucial factor in the technological orientation of the STP

---

[4] If not indicated otherwise, data concerning Berlin-Adlershof is taken from www.adlershof.de (October, 2007).

Adlershof in the 1990s. Around 250 new companies were founded, more than one hundred of them by former Academy employees. In fact, eight of the twelve research institutes located at Adlershof (financed by the City of Berlin and the Federal Republic of Germany) directly descended from the former GDR Academy of Sciences.

A second advantage with regard to the acquisition of qualified personnel was the relocation of the mathematics and natural science faculties of the Humboldt University of Berlin to Adlershof. The decision made by the university at the end of 1991, relocation of the departments of Computer Science, Mathematics, Chemistry, Physics, Geography and Psychology was realized between 1998 and 2003.

3.2 Obtaining the support of a powerful and stable partner in the local or national economic/political environment

As already described in the historical overview on Berlin-Adlershof, the decision to launch the STP project was made by the Berlin Senate in 1992. This decision gave the project political as well as economic importance as it was highly subsidized and received strong political support.

Simultaneously, the wider surroundings of the STP Adlershof were also subject to urban development projects. The airport Berlin-Schönefeld, a 15-minute-ride by car from the STP Adlershof, was chosen to be the only Berlin airport operating from 2010. Until now, there still are three airports in Berlin; a remnant of the former separation. There will be a highway connection (Autobahn) between central Berlin and the airport, which will also pass Adlershof.

Looking at all of these measures, the development of Berlin-Adlershof plays an important role in the urban development of Berlin. The STP Adlershof will thus be able to rely on the stable and enduring support by the City of Berlin.

2.3.3 Creating a recognizable identity of the STP

The creation of an adequate identity and profile of the STP has been the responsibility of WISTA Management. This includes, firstly, the marketing of Adlershof as a "brand" and public relation activities and, secondly, technology clustering.

As to marketing and PR, measures have been for instance: Adlershof events with media representatives and guest lectures; the STP Adlershof homepage (www.adlershof.de); a bi-monthly free magazine (Adlershof Journal); a joint logo that can be used by any Adlershof company. The measures are evaluated by the STP firms on an annual basis.

As to technology clustering, WISTA Management actively participates in the selection and promotion of particular knowledge and economic sectors in order to define the STP's main foci. We will look at these activities in more detail in the next section.

2.4 From business park development to strategic STPM: technology and urbanity

The story of the STPM of Berlin-Adlershof could be titled: From real-estate management to technology clustering and back to urbanity.

2.4.1 Technology clustering

In 1999, the main criticism directed at the Adlershof STP project was the lack of a coherent technology strategy and the absence of an integrated management (cf. Simons, 2003, p. 86-7). Since 2000, WISTA Management has been facing this challenge. In 2002, a new CEO was installed. In 2006, WISTA Management took over the formerly independent business-incubation centers in order to integrate business-incubation services and thereby influence the STP's identity. The crucial measure, however, was strengthening the sectoral approach, based on four major technology clusters, namely:

- Photonics and Optical Technologies
- Environmental, Bio and Energy Technology
- Material and Microsystems Technology
- Information and Media Technology

The definition of the technology fields roots back to the research fields of the former GDR Academy of Sciences. However, only when the Berlin-Adlershof management recognized its strategic value, technology clustering really paid off.

WISTA Management supports the organizational as well as logistic development of technology clusters. Each technology cluster is now represented and managed by a particular technology center, all of them housed in newly built facilities. The STPM was able to acquire renowned architects to realize these buildings such as Sauerbruch & Hutton who designed the award-winning Center for Photonics and Optical Technologies. Thus the technology clusters are represented both in WISTA management and in the physical structure of the Adlershof facilities.

To provide an example: Photonics and Optical Technologies is the leading technology cluster in Adlershof. This technology cluster comprises 54 companies and research institutions with about 900 employees. Companies located in Adlershof as well as other locations throughout Berlin and its surroundings founded the association "OpTec Berlin-Brandenburg" (OpTecBB) in the year 2000. Their objective has been to create a competence network for optical technologies. They chose the STP Adlershof for their head office. OpTecBB launches cross-company co-operations, holds conferences and develops a regional technology network.

The technology cluster with the highest growth rates, however, is Environmental, Bio and Energy Technology. A special focus lies on the field of solar energy and photovoltaics. The STP Berlin-Adlershof has become home to several research institutions and companies in this sector, as shows the relocation of SOLON corporation to Berlin-Adlershof. The company produces solar modules and can be considered as one of leading international manufacturers in its field. The relocation of SOLON with its 220 employees to the STP Adlershof can be regarded as an extremely important milestone for the STPM as it strengthens its profile and expertise in this sector as well as taking it a step further towards a "solar campus" Adlershof.

To sum up the advantages of the Adlershof STPM efforts in technology clustering:

-       Firstly, the internal organization of the STP Adlershof has been improved.
-       Secondly, the profile of the STP Adlershof has been sharpened creating a stronger identity.
-       Last, but not least, the further advancement of certain technologies can be fostered through technology-focused partnerships between private enterprises and research institutions.

2.4.2 Urbanity management

The development of the STP Berlin-Adlershof was not planned as an isolated project to promote science and economy but as part of an integrated urban development project. Thus it contained several areas with different functions: the STP, the Media City, a landscape park as well as residential areas adjacent to the landscape park. The overall plan was for Adlershof to become a "City within a City," an integrated city in the tradition of the European cities (cf. BAAG, 1993).

The reasons for promoting this integrated approach were twofold: Firstly, in the early 1990s the Berlin government expected a large population increase due to the German reunification and therefore a strong demand for residential housing. The demographic forecasts estimated a population growth in Berlin from 3.3 million inhabitants in the early 1990s to 5-6 million at the turn of the millennium. These forecasts turned out to be too optimistic, Berlin's population remained stable. The second reason is the new "urban ideal" of areas with mixed functions. After a long period of pursuing the ideal of spatial separation of living functions in the second half of the 20th century, the mixture of functions has now become the preferred urban planning approach.

Nevertheless, the development of the urban fabric of Adlershof and its surroundings took place very slowly. First steps at the beginning of the 1990s were the removal of old buildings and the conditioning of polluted and contaminated areas. Subsequently, infrastructure such as streets was renewed, the landscape park was developed and first commercial facilities were built. Since 2003, a new residential area close to the landscape park has been developed. But primarily Adlershof has remained a place of work. Today, thousands of employees of the STP and the media city as well as students from the university surge into Adlershof every morning and leave it in the afternoon. Today's commercial and social infrastructure has grown to an acceptable quantity but the hope for a balanced mixture of functions has not been achieved yet.

Measures concerning the social needs and requirement of employees and other agents working in Adlershof are becoming increasingly more important due to the ongoing growth of the STP. Therefore, the future tasks of successful STPM in Adlershof will be a constant and continuous focus on these "weak" location factors. However, Berlin-Adlershof can rely on another advantage with regard to the urban approach. The STP might not be able to create a "real" urban environment in the STP area itself but Adlershof is already integrated within the urban fabric of the city of Berlin. Therefore, Adlershof's management will have to decide which urban features they have to provide within the STP.

2.4.3 The "Adlershof formula"
In summary, the success story of the STP Berlin-Adlershof so far consists of three elements:

(1)   an integrated plan and project of urban development;

(2)   people (from the GDR Academy of Sciences) with an Adlershof identity and the strong will to

develop the location;

(3)   a professional management that operates strategically.

This has been the storyline thus far. As we have argued, the future will depend on technology clustering and urbanity management.

3. Discussion and Conclusion

To conclude, we would like to discuss two points related both to the managerial paradigm of STPM and the evolution of STPs: the professionalization of STPM and the implications for the cluster concept.

3.1 Professionalization of STPM
The increased professionalization of WISTA Management was considered as the main turning point in the history of the Adlershof project. Professionalization has individual and occupational aspects (cf. Mieg, in press). At the individual level (managers, enterprises), professionalization means that work is done according to certain quality standards, including a reliable efficiency. At the occupational level, professionalization follows a certain pathway that holds for any new field of work. In particular, professionalization implies

- the establishment of professional standards and performance criteria and
- the establishment of an international community with some form of organization.

These aspects can also be observed in Berlin-Adlershof. Besides increased global competition, professionalization is another reason for the international diffusion of standards for STPM and, hence, the new role of technology clustering and urbanity management.

As individual professionalization is generally embedded in a context of an already professionalized field of work, it would perhaps have been too early to request professionalized work by the Adlershof STPM from the very beginnings. For instance, the ASTP, the Association of European Science and Technology Transfer Professionals, was not founded until 1999.

Thus Adlershof provides not only a case of a newly developed STP but also an example of a newly developed sector of work, STPM. STPM is still subject to ongoing professionalization both at an individual and the occupational level.

3.2 Clusters

The second point refers to the implications for the cluster concept. STPs represent a special and advanced form of economic clusters. Clusters can be defined as networks of companies in similar or related fields of economic activity, which co-operate closely. In order to do so, a critical mass of participating companies is needed which have to be located in a certain spatial proximity for a certain period of time.

The concept of economic clusters historically developed from the agricultural model of von Thünen's rings in the 19th century (Thünen, 1842) through the concept of industrial districts developed by Marshall (Marshall, 1890) to Porter's economic concept of clusters of the 1980s (Porter, 1990). Von Thünen's agricultural concept depended on the specific distance of a production site to the potential market. Marshall's model was mainly used for the analysis of heavy industries such as steel and automobile industries - which also depend on access and proximity to transport means for their resources and products. Porter's approach emphasizes the aspects of competitive advantages and ongoing innovations in specialized industries due to local co-operation between companies in related or similar industries. Today's discussion on clusters normally refers to Porter's understanding of cluster as business networks.

However, technology clustering, as described in the case of WISTA Management, changes the basic idea of clusters as business networks. The emphasis now is not on business integration but on technology development. This has also been demonstrated in the case of OpTecBB, the regional competence network in optics that connects companies and research institutions in the two neighboring German states of Berlin and Brandenburg. Moreover, in the last 5 to 10 years a large number of similar technology-based competence networks have been founded all over Germany that are now starting to exchange experiences on several platforms (e.g. www.kompetenznetze.de) - in order to learn from one another. Here again, we observe an ongoing professionalization of management tasks such as the definition of technology portfolios, organization of regional conferences, and business incubation.

Technology clustering is only one strategy of STPM, urbanity management is another. The French concept of *technopoles* (cf. Jalabert & Thouzellier, 1990) already integrated the development of STPs with an urban development perspective. However, it placed management in the hands of a public administration and not a professional STPM. The future will show whether even urban development will become a general objective of any professional cluster management. This shift towards technology clustering and urbanity management is what we can observe already *today* in the case of the STP Berlin-Adlershof.

References

BAAG Berlin Adlershof Aufbaugesellschaft Adlershof (Ed.), 2003. Stadt Struktur: Von der Vision zur Realität. City structure: From vision to reality. BAAG, Berlin.

Cabral, R., 1998. Refining the Cabral-Dahab Science Park Management Paradigm. In: International Journal of Technology Management, Vol. 16, No. 8, 813-818.

da Silva, D., 2008. A City and its Science Park: Building a Local System of Innovation for Urban and Economic Development. Paper presented at the XXV IASP World Conference on Science & Technology Parks, Johannesburg (14-17 September).

Etzkowitz, H. & Leydesdorff, L., 2000. The dynamics of innovation: from National Systems and "Mode 2" to a Triple Helix of university-industry-government relations. Research Policy, Vol. 29, No. 2, 109-123.

Fuller, D., 2008. Innovation Campus - the Transformation of a City. Paper presented at the XXV IASP World Conference on Science & Technology Parks, Johannesburg (14-17 September).

Herfert, G., 2005. Berlin Adlershof. Ein neuer ökonomischer Pol in der inneren Peripherie der Stadtregion Berlin In: Beiträge zur Regionalen Geographie, Vol. 61, 208-219.

IASP International Association of Science Parks 2008, Commercialising Science & Technology Parks. IASP, Málaga, Spain.

Jalabert, G. & Thouzellier, C. (Eds.). 1990. Villes et technopoles. Toulouse: Presses universitaires du Mirail.

Marshall, A., 1890. Principles of Economics. Macmillan and Co., London.

Mieg, H. A., in press. Professionalisation. In: Rauner, F, Maclean, R. (Eds.), Handbook of Vocational Education Research. Springer, Dordrecht.

Mieg, H. A., 2006. The sustainability principle as a facilitator of innovative urban development: The case of Berlin-Adlershof. Paper presented at the German-Sino Heidelberg Symposium on Sustainability Oriented Management on the Local/Regional Scale (9-11 March).

Porter, M. E., 1990. The Competitive Advantage of Nations. The Free Press, New York.

Sanz, L., 2008. Impact measures for STPs. Presentation at the XXV IASP World Conference on Science & Technology Parks, Johannesburg (14-17 September).

Siebel, W. (Ed.), 2004. Die europäische Stadt. Suhrkamp, Frankfurt.

Simons, K., 2003. Politische Steuerung großer Projekte. Leske + Budrich, Opladen.

Thünen von, J. H., 1842. Der isolierte Staat in Beziehung auf Landwirtschaft und Nationalökonomie, 2. vermehrte und verbesserte Auflage. Leopold, Rostock.

Fig. 1: Model and current situation of Berlin-Adlershof (Source: www.adlershof.de, the descriptions are added; © Wista Management GmbH)

174

Fig. 2: Development of number of companies and employees in the STP Adlershof 1995-2006
(Source: Annual reports of Wista Management GmbH, cf. www.adlershof.de).

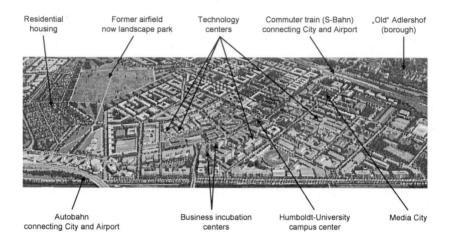

| Residential housing | Former airfield now landscape park | Technology centers | Commuter train (S-Bahn) connecting City and Airport | „Old" Adlershof (borough) |

Autobahn connecting City and Airport   Business incubation centers   Humboldt-University campus center   Media City

Fig. 1: Model and current situation of Berlin-Adlershof (Source: www.adlershof.de, the descriptions are added; © Wista Management GmbH)

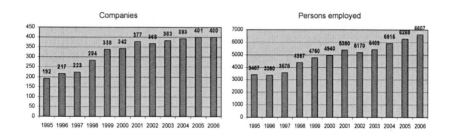

Fig. 2: Development of number of companies and employees in the STP Adlershof 1995-2006 (Source: Annual reports of Wista Management GmbH, cf. www.adlershof.de).

The Research of Eco-industrial Park for Chinese Liquor Production

Shen Peng*, Sun Qihong, Mao Yuru, and Li Yanping
(Key Joint Laboratory on Eco-Industry of SEPA,
Chinese Research Academy of Environmental Sciences, Beijing 100012, China)

Abstract: Liquor industry is a backbone industry in China's national economy. Production enterprises are located all over the country. The analysis of the production status of China's liquor making enterprises summarizes the existing environmental problems. The research is based upon an eco-industry park model with liquor-making manufacturers as major corporations and Yellow Water using enterprises, rice husk silica combination enterprises and ecological agricultural production. We hope the present research establishes a reference for liquor making enterprises in eco-industrial parks (EIP).

Keywords: Liquor making, energy, environment, Eco-industrial Park

* SHEN Peng, Engineer, Chinese Research Academy of Environmental Sciences.

1 Introduction

Liquors in China have a history of several thousand years, and the liquor-making industry is a backbone industry in some chinese provinces. From 1997 to 2004, due to the changes of state's industrial policy in food consumption, liquor output has declined each year, from the maximum of 8.013 million tons down to the minimum of 3.1168 million tons; in 2005 it began to rise. Currently, the output stabilized at 4 million tons.[1]

2 Current status and main problems

2.1 Status

Liquor is mainly produced in the Chinese provinces Shandong, Sichuan, Henan, Liaoning, Jiangsu and Anhui. The products include more than 10 types of flavor, and Luzhou spirits is the largest liquor production, accounting for 70% of China's total output of the liquor.

China's liquor production processes - the example of Luzhou spirit – mainly include: starter making, pit mud production, fermentation, distillation, blending, and packaging, etc.. The main waste during liquor brewing includes: Yellow Water and distiller's grain.

2.2 The main problems

2.2.1 Small-scale enterprises

In 2005, there were 8472 liquor manufacturers, including only about 1,000 large-scale enterprises. But their output accounted for more than 80% of total output. A big gap existed between large-scale enterprises and small-scale enterprises in the material and energy consumption, for instant: water consumption for making one tone 65 degrees liquor is 16 tones in big-scale enterprises and 70 tones in small-scale enterprises. [2]

Another problem regarding small-scale enterprises is that the additional expenses through recycling are difficult to implement. This is why the waste will be treated as cheap feed. Even worse is that these waste may be directly discharged into the environment.

2.2.2 Energy consumption

Liquor brewing takes place in labor-intensive enterprises with manual production process, great labor intensity, poor working conditions, poor business equipment, backward technology and low labor productivity. Luzhou-brewed liquor enterprises in China's Sichuan Province as an example, because of the high level of automated production equipment in its headquarters, shows that the average coal consumption to produce one tone standard 65 is 985.34 kilograms in 2006, but the coal consumption is 3010 kilograms per ton standard 65° liquor in his embranchment by the low level of automation. So there is 3.05 times difference between two working conditions. [2]

2.2.3 Serious Environmental pollution

The water resources pollution in China's Light Industrial is quite serious. The fermentation liquor industry, including liquor brewing, is second to the extent of pollution on the environment and the paper industry ranks first. From the production process, the traditional liquor brewed, using solid state fermentation technology, the main pollutants are fermentation wastewater (or Yellow Water), cooling water, space cleaning water, rinser cleaning water, which Yellow Water and space cleaning water mixed with a large number of natural organic matters, the high COD, BOD, SS concentration exist. Solid distillation, it is necessary to use a lot of cooling water and other wastewater together so that COD and SS concentration could be reduced, but increase the wastewater treatment capacity and processing difficulties. [3]

3 Eco-industry principle

An eco-industrial park is a community of manufacturing and service businesses located [together] on a common land. The members of businesses want to reach enhanced environmental, economic, and social performance through management cooperation of environmental and resource issues. By working together, the community of businesses seeks a collective advantage that is greater than the individual advantage of each company. They only optimize their individual performance for that.

The goal of an EIP is to improve the economic performance of the companies that participate while minimizing their environmental impacts. Components of this approach include green design of park infrastructure and plants; cleaner production, pollution prevention; energy efficiency; and inter-company partnering. An EIP also seeks advantage for surrounding communities to assure that the impact they give is positive.

4 Eco-industrial development major initiatives in liquor making

Since 1993, SEPA has carried out cleaner production and Eco-industry actions to improve economic efficiency, reduce pollution emissions, alleviate the burden of end-of-pipe, improve the product quality and promote sustainable and rapid development. The economic and social benefits were gained.

Based on the Eco-industry and cleaner production principles, the study summarizes the use of waste in the following ways:

4.1 The comprehensive utilization of Yellow Water

Yellow Water is a by-product from liquor-making process, which contains alcohols, acids, aldehydes, esters and other substances, but as well a lot of useful microorganisms, carbohydrate, and small amounts of nitrogen compounds such as tannin and pigment. In recent years, China has done many researches on the Yellow Water reuse; and gets some mature and integrated technologies.

4.1.1 Functional blending liquid producing

Yellow Water contains a lot of acids, alcohols, esters, aldehydes and other materials that can dissolve in water, is soluble in alcohol, or miscible with water and ethanol. Multi-component mixture and formed ethanol-water azeotrope lower the boiling point. It can be distilled from them on 100°C [4]. Through rough filtration, introduce the Yellow Water into distillation columns, using a built-in steam heating. Through this special heating distillation, the Yellow Water, 70% concentration, could be collected as functional blending liquid. After years of storage, the liquid can be directly applied to the middle and lower ends of liquor mixing flavoring. So that not only raises the utilization rate of water, but also reduces environmental pollution.

4.1.2 Vinegar Production

Through filtering the Yellow Water, the suspended solids were removed from it. Acidity of the filtrate and starch are measured in adjusting its alcohol to 7%, then joined Liquor fermentation, and then re-filtered, adjust pH to 8, 5% acetic acid content, adding salt and monosodium glutamate, another sterilization, then packaging. The product, Vinegar, rich in the production of organic acids, amino acids and low lipid material, the quality of products meets the national vinegar standard. It is not only an effective way to solve the water pollution problem, but also increase the economic efficiency of enterprises.

4.2 Comprehensive Utilization of rice husk

Silica coatings are widely used for pesticides, feed, inks, adhesives and other industrial products, resulting in a great demand. Rice husk is the solid waste from liquor brewing process and contains 15-25% SO_2. It can be used to produce silica and try to achieve environmental improvement and economic development of technology production. Manufacturing process include husks hydrochloric acid etching solution, distillation, washing, drying, then after 2-3 hours at 700 ° C, a white powder could be present, it is rice husk silica (RHS). After further processing, silica coating could be produced. Steam generated in the process of burning rice husk can be used to steam liquor, saving coal. [5]

4.3 The development of eco-agriculture

Liquor is a typical "regional resources" product, and its quality depends on the quality of water, soil, climate, air, and other uncontrollable factors, such as human activities, because the microorganism used in liquor-making process have to grow under certain temperature and humidity conditions, and the plants can grow, create, and keep the physical condition. Therefore, massive planting of crops and plants around the liquor production facilities can supply raw materials for the production of liquor, and also improve local farmers income.

4.4 Other Clean Production measures

The main pollutants in the liquor production process are various wastewaters, therefore, based on the concept of cleaner production and its current status, the main measures include the following:

4.4.1 Cooling Water

Cooling water is mainly used to cool liquor steam indirectly for condensation of the liquor steam from gas to liquid. In the conventional production process, the cooling water was taken away from the part of the condenser as wastewater after the heat exchange and drain to the sewage.

At present, some domestic liquor-making enterprises select the used cooling water by the closed-wide recovery network and distribute to bathroom and packaging workshop rinser, and the waste water from bathroom and rinser could be treated and be used as fire protection, irrigation sites, cleaning toilets, turf pouring, and boiler dust removing, etc. Another part of the cooling water is treated as supplementary boiler water, affluent parts of the cooling water re-enter the water pipe network, again for cooling.

4.4.2 Cleaning water

The amount of Cleaned water mixed with abundant natural organic matter, wastewater COD and SS concentration is high and increases the difficulty of wastewater treatment. From the perspective of cleaner production, the production materials and should be filtered out first to decrease the waste water treatment pressure and reduce cost; meanwhile, through strengthened internal management workshop and reduction of material tossed pickling, the COD load also could be reduced.

4.4.3 Cleaning bottle water

Cooling water could be used to clean bottles, increase water recycling in rinser at the same time. It is a measure with a very significant effect that reduces water consumption by over 60%.

5 Liquor eco-industrial park chain network model

Based on the analysis of the production status of China's liquor-making enterprises, summarizing the existing environmental problems and according to the eco-industry and cleaner production concept, an eco-industry park model is carried out with liquor-making manufacturers as major corporations and Yellow Water using enterprises, rice husk silica combination enterprises and ecological agricultural production.

Through the various industries of raw materials, products, by-products and other exchanges, the industrial park industry chain network model proposes symbiotic relations, and enterprise resources can be used for recycling products. The following chart is based on the liquor manufacturers of the eco-industrial park chain network model diagram:

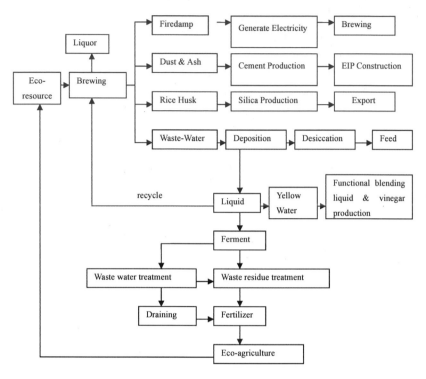

Chart 1. Liquor eco-industrial park chain network model

6 Conclusions

Liquor eco-industry in China should actively promote the concept of sustainable development under the guidance of extending the industrial chain, and vigorously carrying out the re-use of waste. Futher, an eco-industrial park should be established with liquor production as the central industry, and Yellow Water processing, rice husks comprehensive utilization and ecological agriculture as surrounding industries. Like this, the new economic growth points could present and achieve complementarity between the various industries.

Reference:

[1] WANG Yan-cai, the Association of Liquor report in 2005[J], Liquor-making Science & Technology. 2006, (5): 17-24.

[2] GAO Can-zhu. Cleaner Production Standard -Liquor industry (Writing Manual) [R], 2006.11.

[3] ZHONG Yu-ye, SONG Jie-shu. Sanitary Production of Liquors[J]. Liquor-making Science & Technology. 2003, (6): 105-106.

[4] YAO Yu-ying. Chemical Engineering [M]. Tianjin: Tianjin Science and Technology Publishing House, 1993.

[5] JIANG Hong, CHEN Yuan-zhao, ZHANG Liang, ZHANG Su-yi, HU Cheng. Study on Manufacturing Active Carbon from Distiller's Grains[J]. Liquor-making Science & Technology. 2006, (3): 97-101.

Circular Economy in TEDA

Wei Hongmei

Environmental Protection Bureau of TEDA, Senior Engineer

Abstract: TEDA is established as a national demonstration of circular economy and eco-industrial park. TEDA made up a circular economy plan and an eco-park plan, set up a respective organization and an information platform, and made preferential policies to encourage companies in the park to save energy and resources, in order to reduce polluting emission. In addition, TEDA made efforts to develop environmental industry such as recycling of waste electronic products, waste cars, wastewater and recovering electric energy from burning municipal solid waste. So the structure of "promoting by government, enterprises as principal part, public joining" to develop circular economy is under construction in TEDA.

Key words: TEDA, circular economy, eco-industrial park

1. Introduction

This paper investigates and summarizes the state of development of circular economy in Tianjin Economic-Technological Development Area (TEDA), a typical state-class development area. It also gives a brief analysis and review.

Circular economy is a new economy growth mode according to the law of China's economic development and sustainable development. At the operational level, the 3R principles of circular economy (reduce, reuse and recycle) are as follows. The purpose of reducing is to reduce the resource consumption of the economic system, to reduce the burden and costs to the environment and to increase the amount of useful products that resources are converted into. It is a concept of efficiency. The purpose of reusing is to reduce resource consumption by extending the product's life cycle. It will directly improve the efficiency of resources. It is also a concept of efficiency. The purpose of recycling is to substitute the primary raw material resources (especially non-renewable resources) with the new resources converted by the waste products. On the one hand, it reduces the over-exploitation of natural resources by the human beings. On the other hand, it enhances the efficiency of resources utilization by extending the life cycle of the primary raw materials. It is a concept of efficiency, too. So, the core objective of the 3R principles is to improve resource efficiency. Based on the above understanding, implementation of circular economy may include: the ways of reducing-optimization of the industrial layout, adjustment of the industrial structure, and diversification of the product's function, etc.; the ways to make reusable, the components standardized, universalized, easy to disassemble like toy bricks (eco-design), reducing the production of one-time products, developing the remanufacturing industry, recycling of cooling water, producing new durable material etc.; the ways of recycling-utilization of purified wastewater, renewable materials (venous industry) and renewable energy etc.

Circular economy shows the following features. All the elements in the economy system are expedited, flowing, vigorous and circulating. The concept of circulation should be understood from the more extensive spatial scale and time span. Circular economy has its different requirements from tradition. In circular economy mode, the earth resources and environmental capacities are considered as limited and can be exhausted. Exploitation and utilization of resources should be controlled within the appropriate scope to avoid. But it is also market economy. Cost, efficiency and benefits are important and concerned. Developing circular economy does not mean to lower the quality of life. In circular economy mode, the humankind has the ability to use the least resources consumption and environment costs to get economy growth and satisfy the human's demands. The core of circular economy is the unity of efficiency and benefit. Circulation (the process of "resources—products—renewable resources") is an important method to obtain economy growth. By using material circularly, the problem of sustainable use of resources (including environmental resources) can be solved. Circulation is also the ultimate goal of economy development. Circular economy is in coincidence with the laws of economy development. A mechanism should be created to promote circular economy development with laws, regulations and policies. Making resource costs rationalized and environmental costs internalized, with the support of scientific and technological innovation and capital market, will help China to span from traditional economy to circular one.

Currently, there are 3 main levels of circular economy's development in China: Enterprises (cleaner production), Industrial parks (eco-industrial park) and society (circular society).

2. TEDA Overview

The establishment of TEDA was given governmental approval in 1984. It is one of China's first state-class development areas. It is located in the southeast of Tianjin, 45 kilometers from the city center, about 41 square kilometers. The major economy indicators of TEDA have been ahead of other development zones in the country for many years. In the year of 2006, the GDP, industrial output and financial income of TEDA were 78.056 billion Rmb, 303.016 billion Rmb and 18.069

billion Rmb respectively. The four pillar industries, electronics communications industry, machinery manufacturing industry, biomedicine industry and food and beverage industry, completed a total industrial output of 281.94 billion Rmb, accounted for 93.0% of the regional total industrial output. Currently, TEDA has become an integrated region with industry, commerce and life. The city function increases day by day and social undertakings improve comprehensively. In the year of 2006, the permanent population reached 116,300 and employees reached 315,700. During the process of economy development, TEDA always considers environment seriously. Environmental protection is put into regional development planning continuously. The annual environmental investment of TEDA is of over 2% of its GDP. As a result, the environmental quality in TEDA has been good and stable as its rapid economic development.

3. Practices of circular economy in TEDA

The circular economy of TEDA has been going with the global environmental management and the country's economic and social development demand. TEDA was named ISO14000 national demonstration area by the State Environmental Protection Administration of China and environmental management pilot of China's industrial park by UNEP. TEDA became the first demonstration area in April 2004. TEDA became the first "national circular economy pilot" by the National Development and Reform Commission and other six ministries in October 2005. In the process of learning and innovation, TEDA is on a way of its own characteristics in the development of circular economy. They can be summarized as: to persist in the principles of giving priority to efficiency, optimizing the environment and keeping good benefits, to consider the improvement of the efficiency as the core, to consider the technological progress and innovation as a driving force, to change the mode of economy growth actively, to make intensive efforts to establish an eco-industrial park with intensive management, resource-saving and environmental friendly.

The establishment of TEDA circular economy is mainly through two levels of enterprises and region. The outstanding corporate culture is coincided with the concepts of government's development of circular economy. Enterprises independently carried out a large number of practices at different levels. They promoted the development of circular economy in TEDA by their advanced ideas and actions. For example, Novozymes is the world's largest enzyme manufacturer, it transforms the residue after the fermentation of industrial waste and the activated sludge from sewage treatment into organic fertilizers-Novo. Novo's application in TEDA and the surrounding farmland has played a significant role in improving the saline in the coastal region. Another one, Motorola's "green eco-design" includes all environmental designs of the enterprise. It includes alternatives of toxic and harmful materials, lightening of products, recycling use of antistatic materials and reduction of packaging materials (etc.). At the regional level, the ways of TEDA developing circular economy mainly include:

3.1 Controlling the quality of the project strictly, use energy and resources efficiently

TEDA implements the institute of "one vote to reject by environmental protection administration" strictly in its projects. It prohibits the projects with high energy consumption, high material consumption and high-pollution to site in the area and strictly controls the density of the project investment. Thus, TEDA reaches a higher level in land resources, water resources and energy efficiency. In the year of 2006, the land GDP of TEDA was 1.86 billion Rmb per square kilometer. The density of foreign investment reached 750 dollars per square meters. Energy consumption per 10000 Rmb value-added of industry was 172 kilograms of standard coal and consumption of 634 tons of fresh water. All these are max value in the country.

3.2 Establishing institutions and promoting the circular economy by governmental push, autonomy of enterprise and public participation

According to the development of circular economy at different stages, the management committee has established several institutions that are responsible for the related works' planning, organization, implementation and management: the promotion office of environmental management systems and cleaner production, the construction leading group of eco-industrial Park and the facilitation committee of circular economy.

The circular economy emphasizes the multiple objectives of economic efficiency, resource efficiency and environmental effects. The government is more than an effective governor. We should play the role of public and non-governmental organizations to reach a diversity management. The structure of the government, enterprises and the third sector in the form of cooperation, consultation and partnerships is very suitable for the development of circular economy.

TEDA has made a useful attempt in encouraging the third sector to be involved to the development of circular economy. TEDA has set up the branch of the friends of the green, waste minimization club and promoting center of circular economy. The club is provided with a small amount of financial support by the government and is a voluntary union of a number of enterprises. It reaches the purpose of reducing the raw materials, reducing the generation of waste and saving funds, and get a "win-win-situation" of economy and environment. The promoting center is co-established by the Administrative Committee of TEDA and Nankai University. It difiliates government, enterprises, universities and research institutes in the field of circular economy for communication and cooperation.

3.3 Optimizing industry structure by Industrial Symbiosis

Around the pillar industries, TEDA has selectively invited investment according to the requirement of developing the circular economy and optimized its industry structure based on the ecological industrial chain, products chain and waste chain.

The development zone lays stress on optimizing the industry structure. Among the leading industry, major corporations and major products, it improved the ability to the maximum extent, prolong the industry chain and formulates the community intensive advantages. Vertically, it should do well in the matching work of upstream and downstream firms for leading industry. In the upstream, it should support the corporations to set up research and development centers, and lead the industry promotion technically. In the downstream, it should support the corporations to set up sales center and purchase center etc. On the horizontal it should introduce the production of various raw materials, parts and components, so as to continuously optimize and upgrade the industrial structure.

The reverse logistics (building of waste chain) can ensure the achievement of regional recycling utilization for resources. Around the leading industry, the development zone has brought Taiding (Tianjin) Environmental Technology Co.,Ltd., Electronic circuit boards and electronic waste recycling processing projects in, Tian Jin Toho Lead Recycling CO.,Ltd, using the lead waste and lead-acid batteries from the production of storage battery to produce lead alloy,Tianjin Toyotsu Resource Management Co. Involved in automobile dismantling and scrap recycling, Tianjin Rainbow Hills Cast Iron Co. Ltd., using the scrap of the Toyota project production process to make ingots as die materials. These all lay the foundation of the recycling utilization for resources in the electronic and automobile industry.

At present the four mainstay industries of the development zone has initially completed the embryonic forms of industry symbiosis. Taking the Automobile manufacture for example, including the corporations of Japan and South Korea, large number of automotive components enterprises settled down in the development zone. This shows that the development of automobile industry chain becomes integrated and matured. In addition, the garrisons of Tianjin Rainbow Hills Cast Iron Co. Ltd. and Tianjin Toyotsu Resource Management Co.. President Industrial Company which produce lead acid storage battery for electric vehicle and Tianjin Toho Lead Recycling CO., Ltd together forms a close cycle: lead raw materials electrolytic Lead, lead alloy, Lead-acid batteries,recovery of secondary lead from waste batteries. This is the embryonic form of the cycle: resources, products, wastes, regenerate resources .

3.4 Promoting material recycling by technological innovation, enhancing the resources efficiency

The Tianjin development zone has proper ability of technological innovation and industrialization in seawater desalination, regional water cycle and waste incineration power generation. Currently it has already achieved a sewage processing 10 million t/d , renewable water supply by continuous flow micro filtration of 35,000 t/d, RO 30,000 t/d, and Desalination 10,000 t/d. The development zone has established the first artificial wetland system with reclaimed water as its supplementary water source throughout the country, formed a 3.6-kilometer-long artificial ecological channel of biological purification system and 18-hectare lake landscape. The ecological channels and lake landscape "connect" the TEDA wastewater reclaimed system including waster water treatment and comprehensive utilization seawater desalination into "a closed circle", then the wastewater can be reused many times. This also provides the user groups of Toyota Motor, heat plant etc.. So the water using efficiency has increased significantly. The Shuanggang Garbage Power Plant invested and built by Tianjin Development Zone can treat 1/4 of the domestic refuse of Tianjin. It is the unique state demonstration project in garbage power generation industry of China, and its power generation amount can be 1.2 billion kilowatt-hours per year. The research results of this power plant in the nationalization of key equipment, the treatment and zero effluent technology of waste leachate, the technology of comprehensive utilization ashes and residues which come from the incineration of garbage, the CDM methodology of Waste Incineration Power Generation, all have a value of popularization, and these make the development zone head of the country on the technical development and industrialization of waste incineration power generation.

3.5 Building the information platform of circular economy

TEDA through the electronic government affairs system set up the website of ISO14000 Environmental Management System and eco-park in order to introduce the ideas of ecological industry and circular economy. It also has set up the solid waste resource information network, designed and developed two model blocks which are internet investigation of solid wastes and the information of solid wastes exchange and management. The information network makes the information communication for the exchange and reuse of solid wastes resource more convenient.

3.6 Ecological waste management logo, promoting the market development of waste resources

There are no laws and regulations in our country on the regular management for recovery and reuse of waste resources. In this situation Tianjin development zone has innovatively established a new management mode which is different from the "license" mode. This mode grants the "ecological waste management logo of industrial solid wastes" to the enterprises who carry out the waste management in terms of 3R principles, the waste recovery and recycle enterprises, the waste treatment enterprises in the development zone. Then it can achieve the integral ecological management of the waste resources at three levels from source reduction, comprehensive utilization to harmless disposal, and lead the concentrating wastes transfer and utilization in the enterprises with "logo" at the same time in order to promote the effective management and the market development of waste resources.

4 Conclusions

The circular economy is a new pattern of economy growth which can optimize the resource utilization, minimize the environmental pollution and maximize the economic benefits. TEDA has continually absorbed the advanced ideas during its development process. It has experienced an imperceptible influence from "unconscious" to "conscious", boldly practiced and innovated based on the regional characteristics, and on the road of the circular economy. But the circular economy is a systematic process which takes a long time. There would be lots of difficulties, such as theoretical knowledge, technical innovation, system and mechanism, policy and measures, information construction, labor resources and so on. The difficulties will be solved when the atmosphere of developing circular economy in our country gets better.

References
[1] Pierre Desrochers. Regional development and inter-industry recycling linkages: some historical perspectives .Entrepreneurship & Regional Development, 2002,14, 14 :49~65 .
[2] Heinz Peter Wallner.Towards sustainable development of industry: networking ,complexity and eco-clusters[J] .Journal of Cleaner Production, 1999, (7) :49~58 .
[3] Bj rn Stigson. ARoadto Sustainable Industry: HowtoPromote Resource Efficiency in Companies[M] .D   sseldorf: WBCSD, 2001,

Success Story: Industrial Park

(presentation)

Thomas Fichter

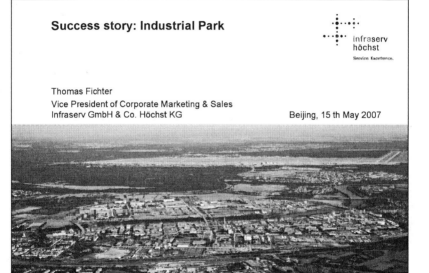

Success story: Industrial Park

infraserv
höchst
Service Excellence.

Thomas Fichter
Vice President of Corporate Marketing & Sales
Infraserv GmbH & Co. Höchst KG   Beijing, 15 th May 2007

Success-Story: Industrial Park

The global chemical and pharmaceutical industry is facing rapid change

infraserv
höchst
Service. Excellence.

Demanding Developments

- More demanding customer requirements
- Continuously increasing market dynamics
- Shorter development cycles
- Emerging new technologies on the brink of industrial application, e.g. nanotechnology, biotechnology and alternative raw materials
- Ongoing globalization and industry consolidation
- Further separation of core business and support functions
- New regulation like REACH

24 07 09   Thomas Fichter   2

Success-Story: Industrial Park

Chemical industry and infrastructure have fundamentally different business models

infraserv
höchst
Service. Excellence.

- Infrastructure is not at the core of the chemical industry business anymore
- The Infrastructure business model is fundamentally different from the typical chemical business models
- Profitability with infrastructure services is lower than targeted profitability in the chemical industry
- The need for profitable growth within the chemical core business triggers the wish to free up capital employed by infrastructure

Criteria	Chemistry	Infrastructure
Technical Innovation	High	Low
Life cycle of plants	10 – 20 years	20 – 50 years
Business model	Spot markets	Long-term service contracts

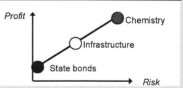

Quelle: CII Group

24 07 09   Thomas Fichter   3

Success-Story: Industrial Park

Chemical manufacturing sites are in a state of transition

infraserv
höchst
Service. Excellence.

Need of Flexible Site Management Solutions

* Portfolio strategies of the large chemical and pharmaceutical companies demand increasingly flexible and innovative solutions

* Transition from commodities towards specialties

* Lower volume products and a generally higher positioning in the value chain require new R&D as well as manufacturing conditions

* Emerging new technologies with strong growth prospects

* Increasing trend for site specialization and consolidation

▶ Large industrial parks as well as niche-specialized industrial parks are most likely to benefit from the ongoing consolidation in the chemical industry

24.07.08   Thomas Richter

Success-Story: Industrial Park

Concentration on core businesses is leading to separation and specialized structures

infraserv
höchst
Service. Excellence.

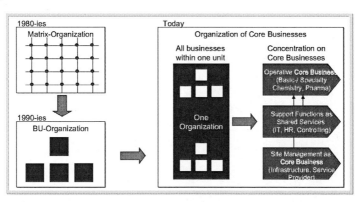

24.07.08   Thomas Richter

Success-Story: Industrial Park

Four distinct site operation models have emerged

infraserv
höchst
Service. Excellence.

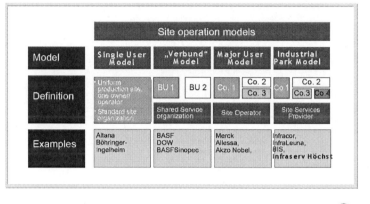

6

Success-Story: Industrial Park

Frankfurt-Höchst Industrial Park: The Chemical and Pharmaceutical Site in Europe's Heartland

infraserv
höchst
Service. Excellence.

7

193

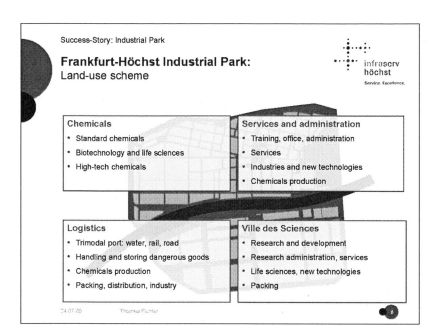

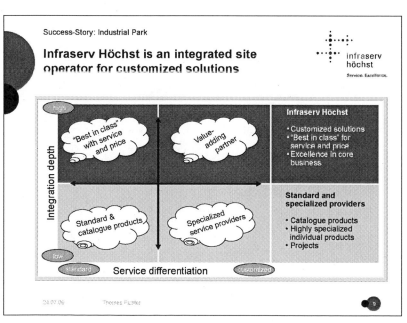

Success-Story: Industrial Park

Frankfurt-Höchst Industrial Park:
Industrial Cluster

infraserv
höchst
Service. Excellence.

Pharma: Global Ranking by Revenues Pfizer: No 1 / Sanofi-Aventis: No 3
Other Global Top Ten Companies: Siemens, Cargill, Sandoz, Hewlett Packard, Air Liquide

24.07.09   Thomas Richter   10

Success-Story: Industrial Park

Infraserv is positioned to be the first choice service partner for Celanese

infraserv
höchst
Service. Excellence.

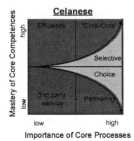

Importance of Core Processes

Infraserv Höchst

* First choice service partner for shareholders and Industrial Park Höchst site companies
* Integrated full service and solution provider for all site services
* Customer value orientation
* Taylor made, customer specific services
* Service level differentiation

Infraserv is the first choice service partner for **Celanese** as full service provider based on it's unique **partner** position and cooperation network

24.07.09   Thomas Richter   11

Success-Story: Industrial Park

Investment in the Frankfurt-Höchst
Industrial Park – Proof of a successful site

infraserv
höchst
Service. Excellence.

in Mio. €

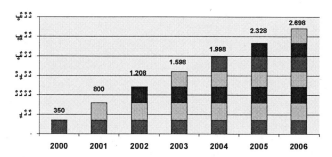

	2000	2001	2002	2003	2004	2005	2006
	350	800	1.208	1.598	1.998	2.328	2.698

Note: HoechstAG invested a maximum of € 256 million in the Frankfurt-Höchst site in individual peak years

24.07.09   Thomas Fichter   12

Success-Story: Industrial Park

Future-ready business
Sustainable Management at Frankfurt-Höchst

infraserv
höchst
Service. Excellence.

At Frankfurt-Höchst Industrial Park you will find one of the world´s most sustainable and attractive business environments in the heart of Europe.

Economic Sustainability	Ecological Sustainability	Social Sustainability
• More than € 15 Bn. GDP contribution	• Compliant to EU and German legislation	• 22.000 Employees working on site
• € 2.7 bn. investment in past 6 years on the site	• Certified auditor for the global environmental management standard (ISO 14001)	• No labor strikes in more than 30 years and counting
• More than € 1.0 bn. infratrade sales volume	• „Zero Regio" EU project lead	• Attractive working and research environment
✓ Economic Bottom-Line	✓ Ecological Bottom-Line	✓ Social Bottom-Line

At Frankfurt-Höchst Industrial Park, companies are not worried about meeting their triple-bottom-line. Why? Because it all adds up – guaranteed!

24.07.09   Thomas Fichter   13

Success-Story: Industrial Park

Thank you very much for your attention

infraserv
höchst

Service. Excellence.

Thomas Fichter
Vice President of Corporate Marketing & Sales
Infraserv GmbH & Co. Höchst KG

Beijing, 15 th May 2007

24.07.08   Thomas Fichter

 14

Success-Story: Industrial Park

Your Questions

infraserv
höchst

Service. Excellence.

24.07.08   Thomas Fichter

 15

Success-Story: Industrial Park

Don´t waste time and money
Let Infraserv handle your waste disposal needs

infraserv
höchst

Service. Excellence.

Our strategy: Your waste disposal ends on site! – Your advantage!

Total security and compliance...

... at the best price.

- **One** of the **largest industrial sewage, waste incinerator and waste water treatment plants** in Germany
- Certified to handle **all non-nuclear wastes** and waste waters
- Waste **disposal center** within industrial site,

  operational 24 hours / 7 days
- Infraserv acting as **official waste disposal representative** for 90% of on-site companies

National Benchmark waste disposal

Source: Internal national benchmark

**Vast experience and high solution competency
in handling any future disposal needs**

24.07.09   Thomas Richter   17

Success-Story: Industrial Park

You like to produce in a clean environment?

infraserv
höchst

Service. Excellence.

You want your plant to be safety and legally compliant.

- Monitoring plant and operational compliance
- Emission and noise control
- Water protection
- Hazard management at Industriepark Höchst
- Radiation protection for plants and facilities
- Waste management
- Monitoring dangerous substances
- Plant safety and occupational safety and health

that is why we handle regulatory matters for you.

24.07.09   Thomas Richter   18

Success-Story: Industrial Park

Industrial Park Höchst, Frankfurt: Peace of mind for investors

infraserv
höchst

Service. Excellence.

- Industrial zone with full legal planning clearance:
  Planning into a secure future in a predictable business environment
- Operation 24 hours on 365 days/year
- Reliable utility systems
- Fast operational permits because of excellent authority relations
- Designed to manage high volumes of raw materials, products and traffic
- Industrial security & emergency response system

24 07 00   Thomas Fichter   20

Success-Story: Industrial Park

Infraserv Höchst Organization

infraserv
höchst

Service. Excellence.

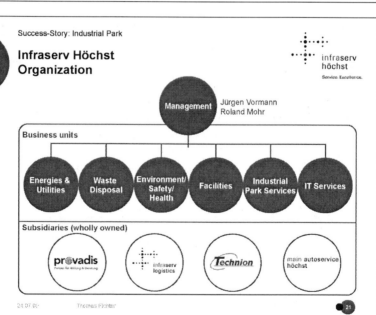

Management — Jürgen Vormann, Roland Mohr

Business units: Energies & Utilities, Waste Disposal, Environment/Safety/Health, Facilities, Industrial Park Services, IT Services

Subsidiaries (wholly owned): provadis, infraserv logistics, Technion, main autoservice höchst

24 07 00   Thomas Fichter   21

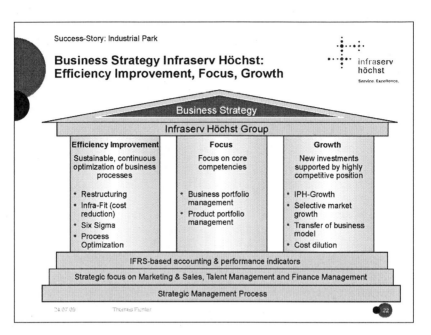

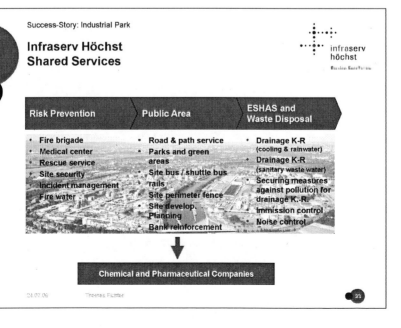

Water for sustainable industrial parks

(presentation)

Emmanuel Duole

Water Management
for sustainable Industrial parks

Did You know...?

- Water is everywhere on earth (about 1385M cubic kilometres)

But

- Only 2.5% is fresh water
  - 90% of fresh water is ice (Antarctic)
  - In the 10 other % only 0.014% can be used as the rest is stored in clouds and ground.
- World water consumption doubles every 20 years
  - More than twice the rate of population growth.

No Water, No Business

- **Industry is the second largest use of water after agriculture**
- **The amount of water used varies widely from one type of industry to another**
  - Cooling water
  - Process water
  - Water for products
  - Water as a medium for waste disposal

🟢 **VEOLIA**
WATER

Cooling Water

- **The largest single use of water by Industry is for cooling in thermal power generation**

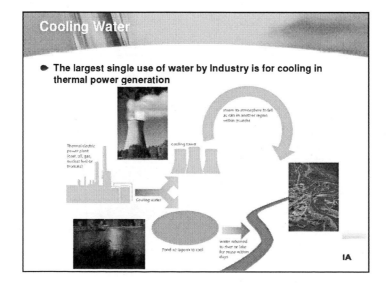

IA

Process Water

- **Industry uses water to make steam for direct drive power and for use in various production processes or chemical reactions**
    - A modern paper mill in Finland has reduced the amount of water used per unit of output by over 90% over the last 20 years: thanks to change from chemical to thermo-mechanical pulp, and installation of a biological wastewater treatment facility that permitted recycling of water
    - A textile firm in India reduced its water consumption by over 80%, by replacing zinc with aluminum in its synthetic fiber production, by reducing race metals in wastewater thereby enabling reuse and by using treated water for irrigation by local farmers
    - A plant converting sugar cane into sugar in Mexico reduced its consumption of water by over 90% by improving housekeeping and segregating sewage from process wastewater

VEOLIA
WATER

Water for Products

- **Food, beverage and pharmaceutical sectors consume water by using it as an ingredient in finished products for human consumption**

Water as a Medium for Waste Disposal

- Many business dispose of wastewater or cleaning water into natural fresh water systems.

- Rivers and lakes can "process" small quantities of waste that can be broken down by nature.

- However, when these limits exceeded, water quality declines and the downstream water is no longer useable without expensive treatment

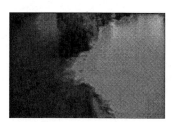

What Can Industry Do to Lighten Water Stress?

- Put its own house in order by
  - Measuring and monitoring water use
    - Understanding the water "footprint" of the business both inside and outside the corporate "fence line"
  - Continuing to reduce water consumption
    - Recycling and reusing water
    - Lowering toxic and other contaminants in all operations involving water
    - Changing production processes to be more water efficient
  - Encouraging suppliers and purchasers up and down the supply chain to adopt best management practices – assisting small and medium sized enterprises to improve water management
  - Innovating
    - Searching for more efficient water treatment technologies

Water Management / Water services

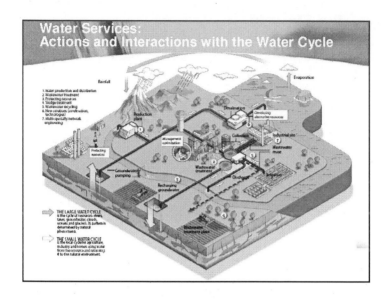

Water Management Strategy for Sustainable Development

Sustainable development refers to development that is economically efficient, socially fair as well as sustainable for the environment on the long term

- **Water is an essential factor for human health, environmental protection as well as economic and social development; water is thus at the heart of sustainable development**

- **Managing water services means facing many challenges:**
  - some are **quantitative**, to fulfill growing demand while preserving the water resource on the long term;
  - others are **qualitative** to insure safe and best quality treatment as a matter of public health and environment preservation.

VEOLIA
WATER

The 3 main axis of water management for sustainability

Water management for sustainability

- Managing the water cycle responsibly

- Placing people at the heart of water services

- Promoting access to water and health

VEOLIA
WATER

Managing the water cycle responsibly

- By protecting water resources
- By optimizing management
- By increasing available water resources
- By involving end-consumers to control water demand

Placing people at the heart of water services

- By ensuring the appropriate training when necessary
- By improving safety and working conditions at all levels
- By transparent communication with stakeholders to improve services and strengthen local attractiveness.

Promoting access to water and health

- By securing high quality and reliable water services
- By promoting optimal use of water

VEOLIA
WATER

Use wastewater services as a tool to preserve ecosystems

- Wastewater services: a central factor to preserve ecosystems
- By controlling all effluents form wastewater systems to preserve and improve local ecological situation.
  - This includes the extension and improvement of wastewater collection networks, the management and treatment of rainwater, high and constant treatment efficiency, etc.
- By insuring the reliability and the performance of the wastewater services, according to international standards.

VEOLIA
WATER

An Example of water management for
Sustainable Development
GUSCO - Thailand

VEOLIA
WATER

Background to GUSCO

- GUSCO (General Utilities Service Company) Established in 1999 to provides total Management of Environmental Services

- GUSCO total current turnover 1,800 million baht (37.5 million Euro/year) from operations activities at 17 sites

- Over 400 employees

- The largest operator of water and wastewater utility services in Thailand

- Core business founded on services to industrial parks

VEOLIA
WATER

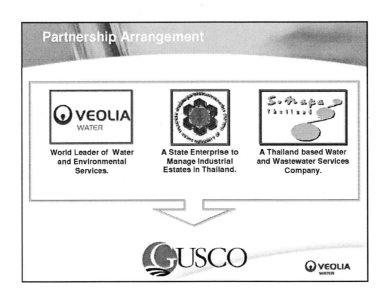

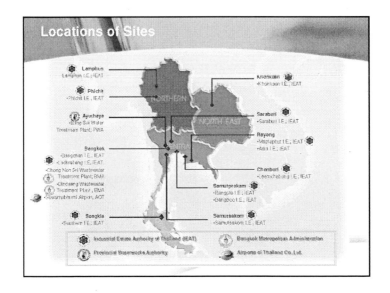

Approach to Sustainability

- As a major Utility Service Provider GUSCO recognizes a social responsibility towards promoting the concept of a sustainable environment

- GUSCO has ISO 14001 at 9 sites and is progressing to cover all major operations locations.

- Local environmental issues that regularly create high profile media interest in Thailand include drought, flood, river pollution, air pollution and the recent Tsunami disaster. The media interest creates pressure to become proactive and to be able to demonstrate actions to address these issues.

- The Laem Chabang wastewater recycle/reuse plant is an action taken by GUSCO together with the building of an learning/educational center at the site. The theme of the learning center is to communicate issues associated with a "sustainable environment"

Thailand – Laem Chabang Industrial Estate

- Key Data Water/Wastewater Treatment

Water Treatment Capacity:	23,000 m3/day
Wastewater Treatment Capacity:	20,500 m3/day
Wastewater Recycle Capacity:	5,000 m3/day
Current Water Demand	22,000 m3/day

Laem Chabang Wastewater Recycle

- **Recycle Plant Drivers**
  - 2005 – Drought in Thailand's eastern seaboard lead to limitations in raw water supplied to GUSCO and consequential restrictions on the treated water supplied to customers.

- **Response & Objectives of Recycle Plant**
  - Primary   - to maximize reuse potential from wastewater.
  - Secondary   - environmental benefits.
    - provide a lead into new markets for Thailand.

- **Key Factors**
  - Water quality and associated production costs.
  - Quality of brine/reject water and its disposal.
  - Competitive pricing in tariff.

Laem Chabang Wastewater Recycle

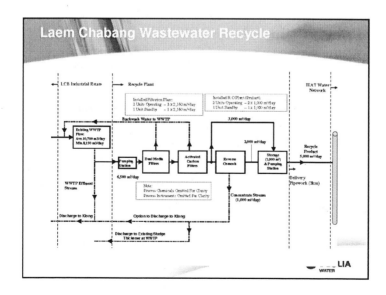

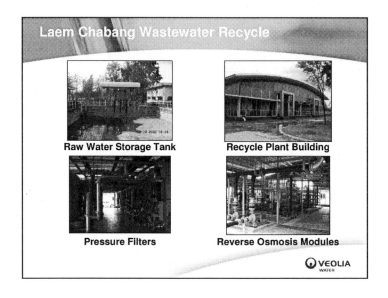

Laem Chabang Wastewater Recycle

Benefits

- Operational
- Environmental
- Commercial

- Opportunities

- Ability to Cope With Short Term Drought
  & Relieve Capacity Problems at Existing   Treatment Plant
- Reuse/Recycle/Optimization of Wastewater
- Entry into Reuse/Recycle Growth Market.
  Client Credibility.
- Use LCB as Showcase for Visitors

Informing and training

- Water management for sustainable development also goes through inform the people and train the staff

- An environmental "Edutainment Center" has been built at the site

  - To inform visitors on the technology used and the operation of the plant
  - to educate on the importance of developing a sustainable environment for the future.

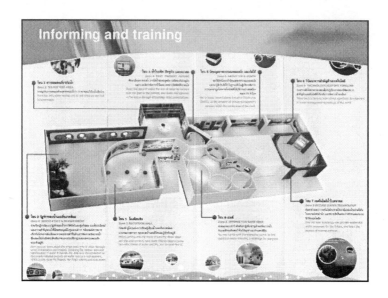

Other Initiatives Associated with Sustainable Environment

Key Performance Targets have been set for individual sites including several associated with environmental issues, i.e.:

- **Water**
  - Targets to reduce water treatment losses
  - Targets to reduce water losses in the distribution network
  - Optimization of water treatment process units and chemical dosing
  - Risk Management measures to reduce risk of pollution
  - Sludge management and disposal

- **Wastewater**
  - Monitoring Individual factory wastewater discharge to the network for compliance with discharge standards
  - Optimization of wastewater treatment process units energy consumption and chemical dosing
  - Sludge management and disposal
  - Risk management measures to reduce risk of pollution
  - Reuse/recycling at 2 sites (Laem Chabang – Operational facility, Lamphun – Stand by facility)

Other Initiatives Associated with Sustainable Environment

- **Water/Wastewater Quality**
  - Continuous improvement in the reliability and efficient optimization of water and wastewater treatment plants

- **Reducing Risk**
  - To identify and instigate potential risk areas to the environment
  - Respond quickly and efficiently to emergency situation

- **Energy Consumption**
  - Continuous improvements to reduce energy consumption and in the application of tariffs

- **Highways**
  - Efficient management of highways and traffic control to minimize accidents, promote efficient traffic movement

Other Initiatives Associated with Sustainable Environment

- **Ground Maintenance**
  - Landscaping and green area conservation to emphasize environmental care and enhance visual impact

- **Communication**
  - Communication and public relations material on environmental issues and the construction of the Laem Chabang Edutainment Center

- **General**
  - An approach to "good housekeeping" for environmental conservation in all aspects of the day to day management at GUSCO Head Office and at site locations.

- **Note:**
  - GUSCO is not responsible for environmental issues associated with air quality, factory emissions, soil protection, waste management and disposal, and pollution control

3.

SUSTAINABLE OPERATION OF
COMPANIES IN INDUSTRIAL PARKS

DEA-based Assessment of Green Cost Efficiency of Ecological Port Public Companies in China

Kuang Haibo

School of management. Dalian University of Technology, China,116023

Abstract: Based on the sustainable development of ecological ports, this article establishes a DEA-based assessment model of cost-efficiency in Chinese public port companies, in which the net value of fixed assets, labor, operating costs and net assets are taken as the input index while equity per share, net profit and prime business revenue as output index, and this model, precisely assesses their green cost efficiency in ecological public port companies. The following conclusion is drawn: in 2004 and 2005, China's port public companies with green cost efficiency took up 50%

Keywords: ecological port, public company, cost efficiency, data envelopment analysis (DEA)

1. Preface

The green cost efficiency of ecological port enterprises refers to an ecological port enterprise's sustainable developing ability to obtain the best output with the lowest cost. It is the measurement of the effective degree of the enterprise's cost minimization and output maximization. Facing the increasingly fierce competition caused by the rapid development of port business and service multiplicity, ecological port enterprises have gradually realized the importance and urgency of improving their efficiency. Thus, it bears great significance to study the cost efficiency of ecological port enterprises.

The cost efficiency research on the ecological port public enterprise mainly includes non-parameter method and parameter one.

As a non-parameter method to study port enterprise's cost efficiency, the Data Envelopment Analysis (DEA)[1] is effective to assess the relative efficiency of the similar economy with multiple-inputs and multiple-outputs using linear programming. Jose Tongzon (2001)[2] once adopted a DEA assessment model to evaluate Australia's port efficiency. Chinese researcher Chen Junfei et al (2004)[3] took tradable shares as input index and equity per share as output index to evaluated 15 ecological ports and shipping enterprises' relative operating efficiency with a DEA model but did not study the cost efficiency. DEA model could assess different indexes without subjective weight and thus enjoy a fairly good objectivity. However it doesn't take random error into consideration [4].

Parameter methods can be represented by SFA (Stochastic Frontier Analysis), TFA (Thick Frontier Analysis) and DFA, which employ the similar principle but differ in diverse presumptions. Guo Hui (2005)[5] used a Bayesian SFA model to assess the efficiency of some container ports both in and outside of China and came to a conclusion that China's large-scale and high technique efficiency container ports had a lower level than foreign ports. Although the methods above have taken random error into consideration, the border functions are of some subjectivity and have considerable influence on the efficiency value [6].

Seiford L.M. et al's (1990) [7] researches lead to a conclusion that the DEA model enjoys a considerable stability in assessing efficiency as well as its suitability to analyze small-sampled efficiency.

Based on the sustainable development of ecological ports, this article establishes a DEA-based assessment model of cost-efficiency in public port companies, in which the net value of fixed assets, labor, operating costs and net assets are taken as input index while equity per share, prime business revenue and aftertax net profit as output index, and this model, precisely assesses the green cost efficiency in ecological public port companies. Data in year 2004 and 2005 of 13 Chinese port public companies are empirically researched.

2. DEA Model

2.1 Introduction

DEA is an effective way to assess the relative efficiency of similar economic activities with multiple-inputs and multiple-outputs with linear programming [1]. Any deviation from the assessment frontier is considered to be of poor efficiency.

Suppose that there are n assessed objects decision making units (DMU), and each DMU has m kinds of inputs and s kinds of outputs. Use X_{ij} to indicate input j of DMU_i, and Y_{ik} indicate input k of DMU_i, and all the inputs of DMU_i could be shown as[11]:

$$X_i = (X_{i1}, X_{i2}, \ldots X_{im})^T, \qquad (i=1,2,\ldots n)$$

The output of DMU_i could be shown as:

$$Y_i = (Y_{i1}, Y_{i2}, \ldots Y_{is})^T, \qquad (i=1,2,\ldots n)$$

The efficiency of DMU_i could be shown as:

$$E_i = \frac{u^T Y_i}{v^T X_i} \tag{1}$$

Where, u^T and v^T are the weight vectors of inputs and outputs. Selected weights u and v properly such that $E_i \leq 1$ i=1,2, ...n.

If we evaluate DMU i_0, which is noted as DMU$_0$, and the inputs are X_0, and the outputs are Y_0, then the relative efficient evaluation model is:

$$Max \quad E_0 = \frac{u^T Y_0}{v^T X_0}$$

$$u \geq 0, \quad v \geq 0 \tag{2}$$

$$\sum_{i=1}^{s} u_i = 1 \text{ , } \quad \sum_{i=1}^{m} v_i = 1$$

Apply the Charness-Cooper transformation and dual theory, and introduce the slack variables s^+, s^- and non-Archimedian infinitely small variable ε. Equivalently transform the fraction programming problem model (2) into linear programming model [1]:

$$\min [\theta - \varepsilon(e_1^T s^- + e_2^T)]$$

$$s.t. \sum_{i=1}^{n} X_i \lambda_i + s^- = \theta X_0$$

$$\sum_{i=1}^{n} Y_i \lambda_i - s^+ = Y_0 \tag{3}$$

$$\lambda_i \geq 0, \quad i=1,2, ...n$$

$$s^- \geq 0, \quad s^+ \geq 0$$

Where, ε is the non-Archimedian infinitely small variable, $e_1^T = (1,1,...1) \in E_m$, $e_2^T = (1,1,...1) \in E_s$, s^- is the vector consisted by the slack variables corresponding to inputs, $s^- = (s_1^-, s_2^-, ... s_m^-)^T$, s^+ is the vectors consisted by the slack variables corresponding to outputs.

2.2 Parameter calculation and result analysis

Parameters λ_i, s^+, s^- and θ can be easily calculated by such software as linear programming. θ is the cost efficiency of an ecological port enterprise. Three results exist. (For detailed information please refer to Tab. 1)

Tab.1 Analysis of DEA model result

Efficient DMU$_0$	When $\theta=1$ and $s^+=s^-=0$, DMU$_0$ is efficient, that is, outputs Y_0 reach the optimization based on the input X_0 among n evaluation objects.
Weak efficient DMU$_0$	When $\theta=1$ and $s^-\neq0$ or $s^+\neq0$, it is called DMU$_0$ is weak efficient. It implicates that an ecological port i_0 could decrease input X_0 by s^- but keep the original output Y_0 fixed, or it could increase the output by s^+ under the input level of X_0.
Inefficient DMU$_0$	When $\theta<1$, DMU$_0$ is inefficient, which means DMU$_0$ could decrease input to θX_0 under original level of output Y_0.

However, in the DEA evaluation model to the relationship with index sign amount of the sample amount of the ecological port, the sample amount is doubled with the index sign amount at least[8]. Therefore have to consider a sample amount relationship with of index sign amount while selecting by examinations index sign.

3 The determination of input and output index in an assessment model of green cost efficiency

Existing researches have done few studies on the input and output of an ecological port enterprise. Reference [3] took assets, tradable shares, labor and operating costs as inputs and equity per share, net profit and prime business revenue as outputs. However, it showed deviation in assessment results due to the lack of some importance indexes such as fixed assets.

Based on the current situation of China's public port companies, this paper selects 3 inputs, namely, net value of fixed assets, operating costs and net assets, as well as 3 outputs, namely, equity per share, prime business revenue and net profit after tax. The total 7 indices chosen in this research can reflect the input and output of ecological port public companies in a scientific and overall way, and meet the requirement of a DEA model that sample of ecological port companies should at least double assessment index.

4 Empirical research

4.1 Empirical data

The empirical data in year 2004 and 2005 of 13 Chinese public port companies are assessed. For convenience, the 13 ecological port public companies is numbered accordingly in table 2. The 7 index original data of 13 ecological public port companies shown in table 3.

222

Tab.2 The N.O. of 13 China Public Port Companies

ecological public companies		No.	ecological public companies		No.
Name	Code		Name	Code	
Shenchiwan A	000022	1	Jingzhou port	600190	8
G Yantian port	000088	2	G Zhongbao	600208	9
Yuefuhua	000507	3	G Chongqing	600279	10
Beihai port	000582	4	G Yingkou port	600317	11
Xiamengangwu	000905	5	Wuhu port	600575	12
Nanjing port	002040	6	G Tianjin port	600717	13
G Shanggang	600018	7			

4.2 Assessment on the green cost efficiency of China's ecological public port companies

4.2.1 Computation of green cost efficiency

Due to DEA model requiring the index data little distinctness and not less than zero, the transformation for the 2004 and 2005 original data of ecological public companies green cost efficiency in table 3 are disposed.

Net assets, net value of fixed assets, operating costs and prime business revenue are divided by 10^9, labor is divided by 10^3, and equity per share is transformed to e power of original data in table 3. The net profit after tax is divided by 10^9 firstly, then it is transformed to e power. The worked input and output data are imported to the model (3). The result of 13 ecological public port companies' cost efficiency is shown in table 4.

Tab.3 Original Data of China Public Port Cost Efficiency Appraisal Indexes

Year	Ecological port pubic companies No.	Input index				Output index		
		Net assets(RMB)	Net value of fixed assets(RMB)	Operating costs (RMB)	Labor	Equity per share (RMB/Share)	Prime business revenue (RMB)	Aftertax net profit (RMB)
2004	1	1852029195	2147623697	499198495	1654	1.08	1552651737	535628927
	2	3528864067	1008839495	332826147	1321	0.537	615920593.4	668498248.4
	3	1012307261	71300666.29	113662064.2	296	0.09	158006365.7	31048847.74
	4	432365010	482781298.9	95798361.03	1345	-0.24	103500590	44949050.43
	5	1135428646	712862321.9	318200402.3	2620	0.33	538619814.3	98214416.98
	6	219202111	199328019	61139279	1120	0.52	195322503	59576992
	7	6794592127	5301218615	345955516.8	3439	0.6405	4043457072	1155744083
	8	1118748000	1769467220	190894122.1	1132	0.126	475347000	119309976.4
	9	620452613.3	946336511.5	739059247.7	1038	0.02	853893989.4	5112298.77
	10	758487301.1	463826933	127036197.9	1723	0.13	206307029.7	29600008.73
	11	1047065610	1523812609	353494084.5	2457	0.36	567196465.3	90174951.75
	12	490220504.9	196673241	63173009.57	1365	0.36	149320127.4	43174926.97
	13	2909828932	1954127524	1023365814	3475	0.2	1886981197	292985445.1
2005	1	2107330402	3192494617	557919974	2031	0.905	1774334469	583452604
	2	3376288318	1238612763	312425672.1	1371	0.55	643577831.2	684480489
	3	1037869426	68575082.57	119838453	279	0.07	161775985.9	25562162.6
	4	436663110.4	494493323	67847827.72	1230	0.02	94822449.49	4298103.07
	5	1211205489	736617657.1	481600719	2703	0.4	748243124.3	117540723
	6	486742940	235486752	64354321	1082	0.39	205365032	60256871
	7	7409843546	8064122700	2276763495	4617	0.6591	4776125938	1189224086

8	1255576000 2068867653 182764413.1	1142	0.143	508028000	135348000
9	618242772 899896082.6 925420267.4	1225	0.028	1113218867	6963263.35
10	756796702.3 485900244.7 165905859	1662	0.1	266494921.4	23066905.2
11	1167842339 2181990885 473008252	2439	0.52	748299866.6	130760199
12	482653460.9 209802781.6 67437170.45	1629	0.34	166633396.5	39805956
13	4014074626 4139338392 1281165241	3653	0.5	2364943473	719262727

Data source:

1.http://disclosure.szse.cn/main/ndbgqw.htm;2.http://www.sse.com.cn/sseportal/webapp/datapresent/SSEQuerySto ckInfoAct?keyword=&reportName=BizCompStockInfoRpt&PRODUCTID=&PRODUCTJP=&PRODUCTNAM E=&CURSOR=1.

4.2.2 Analysis of green cost efficiency results

The figure 1 describes the comparison of ecological public port companies cost efficiency is described. From the table 4 and figure 1, the conclusions could be expressed as:

(1) The following 2004 ecological public companies enjoy an efficiency value of one, Shenchiwan A, Yuefuhua, Nanjing port, G Shanggang, G Zhongbao and G Tianjin port taking up about 50% of the total sample. Wuhu port, Jingzhou port, Xiamengangwu, Beihai port, G Yantian port and G Yingkou port ranked from No. 2 to No. 7, with G Chongqing port ranked the last.

(2) Among the 2005 ecological public port companies, Shenchiwan A, G Yantian port, Yuefuhua, Nanjing port, G Shanggang and G Zhongbao have the best green cost efficiency. Wuhu port, Xiamengangwu, Jingzhou port, G Tianjin, G Yingkou port and G Chongqing port ranked from No.2 to No. 7 and Beihai port ranked the last.

(3) In the successive two years of 2004 and 2005, the green cost efficiency of China's public port companies was stable with the exception of G Yantian port, Beihai port and G Tianjin port, where the green cost efficiency of G Yantian port enjoyed a sharp rise while that of Beihai port and G Tianjin port witnessed an obvious fall. ShenchiwanA, Yuefuhua, Nanjing Port, GShanghai Port and G Zhongbao successively ranked first.Xiamengangwu, Jingzhou Port, G Yingkou port, Wuhu port and G Tianjin port enjoyed a stable ranking in the two years. G Yantian port roared from No. 6 in 2004 to No. 1 in 2005 thanks to the high output/ input level. Beihai port dropped from No. 5 in 2004 to the last in 2005 with an efficiency value of less than 0.5. Table 3 shows that the negative value of output index such as equity per share and net profit is the main reason for the sharp fall of Beihai

port's green cost efficiency. The reason why G Tianjin port dropped from No. 1 in 2004 to No. 5 in 2005 with an obvious fall in efficiency was that the output increased more slowly than input.

The ecological public port companies in China should, based on their specific characteristics, fully investigate and improve the main factors that affect their efficiency in order to raise their comprehensive efficiency.

Tab.4 China' Public Port Companies Cost Appraisal Result

NO.	2004		2005		Two year Average	
	Cost effiency	Rank	Cost effieicy	Rank	Cost effiency	Rank
1	1.0000	①	1.0000	①	1.0000	①
2	0.6341	⑥	1.0000	①	0.8171	⑥
3	1.0000	①	1.0000	①	1.0000	①
4	0.6791	⑤	0.4388	⑧	0.5590	⑨
5	0.7546	④	0.9777	③	0.8662	④
6	1.0000	①	1.0000	①	1.0000	①
7	1.0000	①	1.0000	①	1.0000	①
8	0.7930	③	0.8733	④	0.8332	⑤
9	1.0000	①	1.0000	①	1.0000	①
10	0.5503	⑧	0.6530	⑦	0.6017	⑧
11	0.6061	⑦	0.8083	⑥	0.7072	⑦
12	0.9775	②	0.9786	②	0.9781	②
13	1.0000	①	0.8660	⑤	0.9330	③

5 Conclusion

1) An index system to assess the green cost efficiency of ecological port public companies has been structured, and a DEA-based assessment model of green cost efficiency of ecological public port companies has been established to objectively evaluate the green cost efficiency of ecological public port companies and to make up the lack of an overall assessment and comparison of existing literature in ecological public port companies' green cost efficiency.

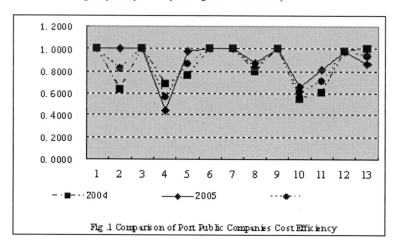

Fig .1 Comparison of Port Public Companies Cost Efficiency

2) Data of 13 China's ecological public port companies (Shenchiwan A and G Shanggang etc.) between 2004 and 2005 are collected and assessed with the DEA-based assessment model of green cost efficiency of ecological port public companies. Then the green cost efficiency value and relative order of those companies are calculated.

3) The results of the empirical researches showed, 1) in 2004 and 2005, approximately 50% of the ecological public port companies have a high green cost efficiency. 2) In 2004 and 2005, the green cost efficiency of China's public port companies was stable with the exception of G Yantian port, Beihai port and G Tianjin port, where the green cost efficiency of G Yantian port enjoyed a sharp rise while that of Beihai port and G Tianjin port witnessed an obvious fall. China's ecological port public companies should improve the factors which affect their cost efficiency.

References

1. Sheng Zhaohan, Zhu Qiao, Wu Guangmou. DEA theory, method and application [M]. Science Pubishing Company. 1996, 2-13, 65-72,155.

2. Jose Tongzon. Efficiency measurement of selected Australian and other international ports using data envelopment analysis [J], *Transportation Research Part A*, 2001, (35): 113-128.

3. Chen Junfei, Xu Changxin, Yan Yixin. Appraisal fo operating efficiency of listed companies in port water transportation based on data envelopment analysis[J].Journal of Shanghai Maritime University, 2004, 25(1):51-55.

4. Kevin Cullinane, Teng-Fei Wang, etc. The technical efficiency of container ports: Comparing data envelopment analysis and stochastic frontier analysis [J]. *Transportation Research Part A*, 2006, (40): 354-374.

5. Guo Hui. Analysis on TE of container port – a comparing of Chinese container ports with their counterparts in the world. Master thesis of Dalian Maritime University, 2005. 3.

6. Kaylee A. Garden, Deborah E. Ralston. The x-efficiency and allocative efficiency effects of credit union mergers Journal of International Financial Markets [J].*Institutions and Money*. 1999, 9, 285–301.

7. Seiford, L.M., Thrall, R.M. Recent developments in DEA. The mathematical programming approach to frontier analysis [J]. *Journal of Econometrics*. 1990, 46, 7-38.

8. Banker RD, Charnes A, Cooper WW. Some models for estimating technological and scale inefficiencies in data envelopment analysis [J]. *Management Science*. 1984, 30 (9), 1078-92.

Research on the Mode of Production Management

Based on Bill of Material Information Share

Tan Chengxu[1] and Luan Qingwei[2]

1 Management School, Dalian University of Technology, Dalian 116024, PR China

2 Information Department of Dalian Government, Dalian 116011, PR China

Abstract: Aiming at a traditional production management mode, information existing in the production process, is difficult to be integrated and shared. Using the Bill of Material (BOM) information sharing as basis, a production management model is set up, which is based on the Pro/I structure's BOM. There are three characters in the model: Firstly, adopting Pro/I systems to integrate AutoCAD software. Secondly, it's applied to complex product assembly and manufacturing process material delivering, helping to confirm BOM's function and performance. Thirdly, it realizes the production-correlated data's unified management. This helps the corporation to enhance their colligation management level and competition ability. By analyzing the examples, some key techniques are laid out, and this exploits a new idea for the corporation's production management mode.

Key words: BOM; ERP, PDM, Pro/I structure, Production management mode

0. Introduction

In the manufacturing industry, the BOM (Bill of material) constitutes a list of the production material that produces a certain product. The data of BOM is used to describe the product material and constitutes the relation between them such as assemble etc. At the same time BOM can indicate other information in a related text file, such as the product manual, the product packing, components and parts etc. [1]. As a kind of basic data, BOM process is the core position in information-based engineering of business enterprises. BOM is the basic form [2] of transferring data among different sections and different processes in the enterprise, and it is the bridge that links the product engineering design and operation management.

Currently, foreign research on BOM information generally aims to imbed the BOM information management in various systems of ERP (Enterprise Resource Planning)[3] or PDM (Product Data

Management)[4], but there is only a little or no other BOM data which can be used in each working section. The PDM software such as Intralink, Iman, etc. [5] manages a great deal of data of the design BOM and the craft BOM concerning the product, but there is still blank about using the BOM information in the production management model, especially the BOM information under Pro/I structure.

Domestic research on BOM information has just started, e.g. "The research of plan system in mixed production" by Liu Mingzhong [6], "The material apply BOM mode according to the craft process" by Xu Jianping and Guo Gang[7], "The cost control research of various BOM products in spinning according to the ERP system" by Zou Xin, Huang Huijun. [8] They concentrate on machine, computer, automation, information and management, etc. Along with the development from MRP [9] to ERP, the BOM expands from the single design BOM in the MRP period to various forms. Different BOM information is carried on the maintenance respectively by the homologous information system. The exchange, share and the search outlet of information resources are not unimpeded between different sections of one department, which is a widespread problem that exists in the production line.

In consideration of the above-mentioned factors, this article makes use of the integrated BOM information sharing as a basis and sets up a production management model which is based on the Pro/I structure's integrated BOM. The research combines the actual production (or rather: development) of a local motorcycle factory, and integrates information of product design, craft, manufacture, sale, etc.

1. The principle of production management mode according to the share of BOM

1.1 Principle of the BOM information share

Aiming at the problem of traditional production mode, the existing research put forward solutions from different angles, such as the conjunctional design, parallel engineering, network's production, etc. [10], and require the information sharing and integration from different design and production stages. Thus the set up of an information management mechanism is the key to improve traditional production modes.

The integrated BOM sharing basically includes each stage. The usage and product of BOM's vision are related with products process; integrated BOM can manage to organically, analyze and extract proper data from different system and business data, then hand over the information to the right position in time.

The first advantage of BOM sharing principle is to analyze and extract proper data from different applied systems and hand it over to the right position in time; second is to achieve the share of information present at in the stages such as design, craft and so on.

1.2 The operation principle of BOM's product information

The whole product management process of an enterprise includes collecting, delivering and processing data; its product is the material performance of data. The output data of BOM is the core of enterprise's information flow; BOM information can be taken as the valid project of mining and integrating data. The design section can use the system of CAD to carry on the general and detailed designs of the product after receiving the order, and form various detail of BOM about the product design; then the craft section can withdraw total detail and classification detail information from the system of CAD, and produce the data information that CAPP system need (realizing the conversion from design BOM to craft BOM); after carrying on the design craft information in system of CAPP, the information can be analyzed, then from craft BOM to manufacturing BOM can be made.

The first advantage of various BOM information sharing is to integrate information in enterprise's process; second is to shift from craft BOM to manufacturing BOM.

1.3 The principle of BOM production management mode

Being described as a kind of structured form for the basic data of each link, BOM is the important connection of design, craft, production, supply, sale sections etc. It is the basis that can ensure the engineering design and craft sections carrying on the CAD/CAPP, the production management section implementing MES/ MRPII/ ERP, and the manufacturing section carrying out the CAM [11][12]. In the life cycle of a product, different sections do the design, management and usage of BOM on own purpose: design BOM organizes data extracted from the product design paper; craft BOM manages craft files; manufacturing BOM organizes manufacture of product; quantity BOM checks product quantity information, discovers and controls key quantity link. The integration of different BOM forms the production management mode of integrated BOM.

First advantage of integrated BOM is to carry out information management mode of each link; second is to form production mode through integration of different BOM.

2. Establishment of the production management mode according to the BOM information sharing under Pro/ I structure

2.1 The integration of BOM information according to the Pro/ I structure

In the product manufacturing process, CAD information of product design is the foundation on which CAPP and CAM systems collect information. Integration of BOM information under Pro/ I structure, with the design software of AutoCAD, collecting information, forms the most basic design BOM data. Based on design BOM, other systems form BOM data that each applied system need through mechanism BOM.

The advantage of BOM information under Pro/ I structure is to design software as AutoCAD through Pro/ I system, collects information that design BOM need and forms the most basic design BOM data in the system of integrated BOM.

2.2 The integrated BOM information mode

Currently, there are three methods of the mechanism of BOM integration: the appropriative BOM management system, PDM/PLM and the ERP system. Enterprises which applied PDM (Pro/ I) product, realized integration among the whole process. We use PDM's seal and pack function to pack CAPP in PDM, carry on the information sharing between CAD and CAPP under PDM. We adopt integrated method of wreath-shut dynamic craft, integrate CAPP with production adjustment system, form the track of wreath-shut, let CAPP do the craft's limited ability design, change and adjust immediately according to dynamic feedback on resources condition and make systems get excellent craft project to guide production. Thus let the PDM system of enterprise become the unified resources management system; design a parallel product development environment. PDM's seal and pack function lead to the integrated BOM as Figure 1.

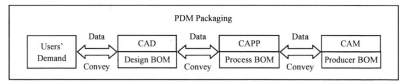

Fig1.   The model information of integration BOM

The first advantage of BOM integration is to integrate three types of BOM information effectively in the production process through the PDM system, unifying all the resources. First, according to the product structure it brings persons to establish, communicate, support the design intention share and deliver all the different structural by sharing and delivering data. The second advantage is to overcome boundary of each section, and to manage the information processes of design → craft → manufacturing and extend the BOM information to the quality checking and purchase section etc. The third advantage is to build up an united resources management system, unify various resources, according to product structure.

2.3 The production management mode of integrated BOM information sharing under Pro/ I structure

From figure 1 we know that the integration of various BOM forms can be carried out through PDM's seal and pack. Taking excelsior design process as an example, we can excavate the Pro/ I function and turn the applied degree of PDM software wider and deeper. In order to realize the faster running of design and the quick conversion of design information.

2.4 The mode's special features and innovation

The special features and innovations of this model are expressed in three aspects at least:

(1) This mode uses integrated BOM information sharing as basic; production management model based on Pro/I structure is set up; it's applied to complex product assembly manufacture process' material deliver, helps to confirm BOM's function and performance, control BOM multi-operation conformity in the production industry.

(2) The integrated BOM information under Pro/ I structure carries out the integration with design software as AutoCAD through Pro/ I system, collects the information that design BOM need from these software and forms the most basic design BOM data in system of integrated BOM.

(3) This research combines product movement physically, integrates product design, craft, manufacture, sale and etc. Effectively, it carries out unified management of product's data, thus raises comprehensive management level and competition ability of enterprise and exploits a new thought for the corporation's production management mode.

3. Application examples

3.1 Basic information

A local motorcycle factory has carried out of Pro/I product. Combining the actual needs of the enterprise, it has integrated effectively various BOM in production process and solved several key techniques in solution project of production management mode of integrated BOM information sharing under Pro/ I structure.

3.2 Assurance for reflects and conversion relation between integrated BOM views

The assurance for reflects and conversion relations between integrated BOM views is the key step that links up each product process. The usages of various BOM in integrated BOM have related with developing process: various BOM is a kind of vision relation on tri-space determined by product type, application realm, and life cycle of product. The conversion from design BOM and craft BOM to manufacturing BOM is the most important, difficult thing and must be completed at first. The conversion of manufacturing BOM is mainly achieved by definition of three kinds of components and parts-fictitious, central and outer. Relevant documentaries have detailed treatise of specific conversion relations [13]. Based on craft BOM and related information, other BOM can be got by carrying on statistics from some aspect information of manufacturing BOM.

3.3 The description of integrated BOM data

From data structure, various BOM diagrams' data structure in the integrated BOM entirely can be divided into three kinds:

(1) The data structure of design BOM and manufacturing BOM which reflect the assembled relations between components and parts. In order to guarantee the integrity, consistency and accuracy of design BOM and manufacturing BOM, we describe them by two relations, the assembled relations of father-son typed components and parts and natural attributes relations of the components and parts. This kind of data structure has the advantage that the components and parts relations and natural attributes relations in great quantities only need to be described once.

(2) The data structure described product processes and assembles of craft BOM. Data structure of craft BOM, must satisfy technique request for BOM diagram conversion in various BOM at the same time. As two relations, "Process procedure of components and parts "and" Produce type of components and parts " are adopted to describe craft BOM.

(3) The data structures are described by BOM, quantity BOM and cost BOM. This kind of BOM data takes craft BOM as the basis, does statistics or discriminates the some aspect's information of manufacturing BOM, tending in simple, and the information is clearer and concentrated. It can adopt descriptions of data area, such as" the material code" and" special realm information" etc.

3.4 The edition management of integrated BOM data

There are differences between the edition management of BOM product data and that of PDM. It involves numerous middle or temporary product data in design process that had not been confirmed and examined; the editions managements of BOM product data involve the product data which had been confirmed and examined, and attempts to describe the edition problems in various BOM data through the approval data structure.

3.5 The comparison of production management model of BOM information share under the Pro/I structure and traditional production management mode

The production management model of BOM information share under Pro/ I structure and the traditional production management model have been compared on four aspects: product-developing period, product cost, product quantity, and link up between different sections (see table 1).

Tab 1. The comparison in product operation of two kinds production management modes

Four aspects of comparison	Traditional production management mode	The production management model of integrated BOM information share under the Pro/ I structure
(1) The product developing period	The product-developing period is long. The product development, particularly the development of new product usually need to modify continuously, thus taking long time for the corporation of different sections.	Product-developing period is short. Product development, particularly the development of new product adopts the integrated various BOM information, have no need of the coordination between different sections.
(2) Product cost	Increase coordination of each section, the extension of product design, craft weave and product organizing period will make the products cost increase.	The decrease of section coordination, the shortened period of product design, craft weave and product organization will make the product cost lower.
(3)Product quantity	Variety product series, increase coordination of different sections make design, craft, manufacturing sections feel heavy of mission and make disadvantageous influence on the produce period and quantity. Thus difficulties in product quantity assurance.	Though there are varieties of product series, usage of new model can decrease coordination between sections, enable design, craft, manufacturing sections to integrate information of various BOM. Thus there have assurance to the manufacturing period and the manufacturing quantities.
(4) The link up between different sections	Taking craft and design for example, the craftwork is an indispensable wreath for every manufacturing enterprise, but under this kind of model, the craft usually becomes the bottleneck in the production because of lacking the information sharing mechanism between designs and craft weave. The link up of section work is difficult.	Different sections adopt the information sharing; taking craft and design for example the craftwork is an indispensable wreath for every manufacturing enterprise. Under this model, information sharing mechanism between designs, craft weave is established. The link up of section work is very easy.

4. Conclusion

(1) On basis of integrated BOM information sharing, a production management model based on Pro/I structure has been set up and applied to complex product assembly manufacture process' material delivering. This helps to confirm the BOM's function and performance, control BOM multi-operation conformity in production industry and resolve problems existing in the traditional production management model such as long product-developing cycles, high product cost, difficulty in assurance of product quantity, difficulty of linking up different sections etc.

(2) In traditional production management mode, the main reason and result of problems is that it can hardly carry out sharing and integration of information between different sections. However, production management with BOM information sharing based on the Pro/I structure is a valid path

to resolve problems of the traditional model. It adopts combination of several BOM, integrates different stages' information of production process, thus speeding product development and raising the management ability to produce in general.

(3) This research combines the product movement physically, integrates the product design, craft, manufacture, sale and effectively, carries out the unified management of the product's related data, thus raising the comprehensive management performance and the ability of the enterprise to compete in the market.

References

[1] Graham Hooley, John Fahy, Tony Cox. Marketing Capabilities and firm performance: a hierarchical Model [J]. Journal of Market-Focused Management, 1999, 4(3): 259-278.

[2] Siu Wai,Sum.Small Firm Marketing in China: a Comparative Study[J].Small Business Economics,2004,12(8)691-726.

[3] WYLIE L.ERP:a vision of the next-generation MRPII[J].Computer Integrated Manufacturing,1990,S-300(339.2):1-5.

[4] MILLER E.PDM today [J]. Computer Aided Engineering, 1995,14(2): 30-45.

[5] Luo YH,et al. A Remote Cooperative Design System Using Interative 3D graphics [J].International. Journal of Image and Graphics,2004,2(4):653-667.

[6] Liu Mingzhong etc., The research of plan system in mixed production, [J]2005,11(4):530-534.

[7] Xu Jianping, Guo Gang, The material apply BOM mode according to the craft process, [J]2005,28(6):19-23.

[8] Zou Xin, Huang Huijun etc., The cost control research of various BOM products in spinning according to the ERP system, [C]2004,11.

[9] Weight Oliver.The Executive's Guide to Successful MRPII[M].USA:Oliver Weight Publications,1992,45-97.

[10] Wang Liyun, Xiao Tianyuan, Yang Nan. The research and realization of developing platform. Computer Integrated Manufacturing System—CIMS, 2002, 8(8): 640-644.

[11] Li Bo. The parallel engineering and DFX technology. CAD/CAM and manufacturing information. 2003, 10: 5-8.

[12] Li Y L,et al. Design and Implementation of A process oriented intelligent collaborative Product'. Design system[J].Computer Industry ,2005,8(6)126-145.

[13] Liu Xiaobing, Huang Xuewen, Ma yue etc. The xBOM research of the product's life cycle. Computer Integrated Manufacturing System—CIMS,2002,8(12): 100-105.

Research on Multi-Layer Fuzzy Synthetic Evaluation Model

of Industrial Group's Synergetic Capacity

Zou Zhiyong and Shang Hua

School of Management, Dalian University of Technology, Dalian, China, 116023

Abstract: The paper begins with the concept of an enterprise group's synergetic capacity. Then it builds up the theory model and the evaluation model of enterprise group's synergetic capacity and calculates the index of enterprise group's synergetic capacity by using SPSS, level analysis methodology and fuzzy integration evaluation methodology. The object of building up the system and the model is to provide the theoretical basis and an instrument to enable enterprise groups to evaluate its own synergetic capacity.

Key words: Synergetic Capacity, Fuzzy Evaluation Model

1. Introduction

From the main paradigm of the enterprise theoretical capacity, it began with Wernerfelt's resource basic theory (1984) as its development starting point, and then was promoted by Prahalad and Hamel (1990), Leonard-Barton (1992) core competencies theory research. Teece, etc. (1997, 2000) the dynamic capacity theoretical study, became this theory's officially formed signs. And then new theoretical development-enterprise knowledge theory [1] emerged. But all these have not yet studied the enterprise synergetic capacity, and have not involved the evaluation of synergetic capacity. But for the enterprise community in reality, especially for the enterprise group which includes dozens or even hundreds of members of corporate enterprises, the operation and management must be a complicated system engineering. And it requires the joint participation and close collaboration of the members of their enterprises, departments and relevant staff. The level of the enterprise group's coordination capacity becomes the key to effectively distribute its internal commodity resources, and that is, to measure enterprise groups' economic performance [2].

2. Building up a model based on the SPSS theory of enterprise groups synergetic capacity

Enterprise Groups are aggregates caused by various factors, the main variety, a variety of synergetic links, and they form a complex system by the large number of units and levels, and various elements of the organic integration [3]. And the synergetic capacity exists universally in the system, which is the proportion of the synergetic factors in the management system of each enterprise, as well as the overall level of coordination among factors [4]; according to certain manner, various factors of the interact, coordinate, synchronize, and produce order parameters which dominate system development. And they dominate system's orderly and steady development, thereby allowing the system to function in the overall doubling or amplification, that is, to achieve "2 +2> 4" synergies [5].

Tab.1.1   Synergetic Factors

Synergetic Factor	Synergetic Factor	Synergetic Factor
Strategic Coordination	Organizations Coordination	Production Coordination
Cultural Coordination	Financial Coordination	Commerce Coordination
Relations Coordination	Resources Coordination	Technical Cooperation
Knowledge Coordination	System Coordination	Marketing Coordination
Brand Coordination	Projects Coordination	Process Coordination
Merger Coordination	Logistics Coordination	Purchasing Coordination
Competition Coordination	Capital Coordination	Leases Coordination
Institutional Coordination	Information Coordination	Innovation Coordination

According to these 24 kinds of synergetic factors, we designed a questionnaire, and survey the EMBA students of Dalian University of Technology since 2003 on the coordination capacity of enterprise groups. Respondents according to their own understanding of the actual situation of enterprise groups, sort and label various factors of each group according to relative importance , "1" means the highest, "2" means followed the same token, "8" means the minimum.

We issued a questionnaire of 138 copies, and have received 126. After that, the criteria were set up to eliminate void questionnaires. The first rule was to assess whether respondents seriously fill out a questionnaire, and the second rule was whether the questionnaire was completed in accordance with the requirements. After removing invalid questionnaires, we got 121 effective questionnaires.

We have a factor analysis on the results of the survey by using SPSS, and then we got the load factor matrix after rotation of Table 1.2.

Tab.1.2 Rotated Component Matrix(a)

	Component		
	1	2	3
Resources Coordination	*.846*	.086	.150
Organizations Coordination	*.791*	.089	-.052
System Coordination	*.756*	.361	-.095
Information Coordination	*.632*	.289	.082
Strategic Coordination	.210	*.753*	-.301
Cultural Coordination	.321	*.681*	.008
Process Coordination	.060	.011	*.844*
Leases Coordination	-.067	-.083	*.809*
Innovation Coordination	-.058	-.077	*.784*
Commerce Coordination	.159	-.012	*.699*

Extraction Method: Principal Component Analysis. Rotation Method: Varimax with Kaiser

Normalization. A Rotation converged in 7 iterations.

From the load factor matrix result of Table 1.2, we can define: the first group of factors is the meso-level coordination factors, which influence each other and form the synergetic capacity of meso-level; the second group of factors is the macro-level synergetic factors synergy, which influence each other and form the synergetic capacity of the macro-level; the third group of factors is the micro-level synergetic factors, which influence each other and form the synergetic capacity of the micro-level; all these three levels influence each other and compose enterprise groups synergetic capacity. Thus, we get enterprise groups theoretical model of synergetic capacity (Figure 1.1 shows).

2.1 The establishment of enterprise group evaluation index system of synergetic capacity

2.1.1 Hierarchical structure model of enterprise group evaluation of synergetic capacity

According to the above theoretical model, we can design an enterprise group evaluation index system of synergetic capacity, which have three levels, showed as Table 2.1. The first class includes 1 indicator, and the second class includes 10 indicators, and the third includes 29 indicators.

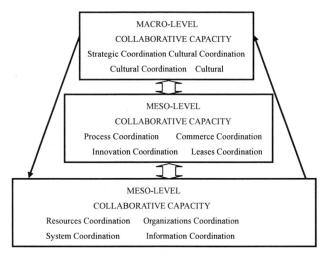

Figure 1.1 Enterprise Groups Theoretical Model of Collaborative Capacity

Table 2.1 Enterprise Group Evaluation Index System of Synergetic Capacity

Level-one Indicator	Level-two Indicator	Level-three Indicator
Enterprise Group Synergetic Capacity valuation U	Strategic Coordination U_1	Member enterprises are able to recognize the importance of strategic coordination U_{11}
		Member enterprises are able to coordinate and implement strategic projects U_{12}
		Do complementary strategies to support and share strategic risks U_{13}
	Cultural Coordination U_2	Group Culture is the carrier of group business philosophy and group awareness, the soup of group U_{21}
		The establishment of common conduct conducing to group coordination U_{22}
	Resources Coordination U_3	During stocking, timely deploy and integrate raw materials according to production plans U_{31}
		Production equipment and technical process can timely deploy and integrate according to demand changes U_{32}
		Human resources can timely be adjusted and trained according to the needs of the development of enterprise group U_{33}
		Founds can be deployed timely according to the company's need U_{34}
	Organizations Coordination U_4	The flat organization structure extent U_{41}
		The situation of sharing group's public organization U_{42}
		The extent of group diversification U_{43}

	System	System takes on powerful implementation U_{51}
	Coordination U_5	Consistency of all companies' rules and regulations U_{52}
		Rationality and practicability of all regulations U_{53}
	Information	Application of information technology in many internet-based companies U_{61}
	Coordination U_6	Enterprise' accumulated knowledge rapidly transfer and circulate in different business units and departments U_{62}
		Focus on information collection, storage, processing, transmission, utilization , and so on U_{63}
	Process	Stock together, launch into market of goods supply U_{71}
	Coordination U_7	Share production capacity, achieve the information exchange of production links with other sectors U_{72}
		Share brand and sales channels, all sides of marketing realize information coordination U_{73}
	Innovation	Jointly established R&D organization U_{81}
	Coordination U_8	Communicate innovation experiences and achievements U_{82}
		Share technology and R&D achievement, exchange and use technology each other U_{83}
	Leases	Normative acts among members to reduce synergetic risk U_{91}
	Coordination U_9	Achieve institutional protection and support U_{92}
	Commerce	Enterprise groups unify internal business language and business platform U_{101}
	Coordination U_{10}	Enterprises can conduct business through the internet activities U_{102}
		Achieve commercial data exchange between enterprises U_{103}

2.1.2 Structure of evaluation model

In this paper, by using multi-storey structure vague comprehensive evaluation method, we construct the evaluation model of enterprise group synergetic capacity [6]. Firstly, we use Analytical Hierarchy Process to build up synergetic multiple indicator systems, and calculate index weight, and then amend the weights, and adopt fuzzy method to evaluate synergetic capacity index[7].

Analytical Hierarchy Process (AHP) weight vector calculation

(1) Each expert judgment matrix-tiered list;

(2) use an eigenvector method to judge matrix weight vector;

(3) by each judge matrix consistency test, to determine their choice;

(4) by a weight vector comprehensive method, judging the weight matrix-vector and according to different people under the same criteria (5) seeking synpaper weight vectors.

Fuzzy comprehensive evaluation

The mathematical model elements of fuzzy comprehensive evaluation :

(1) Factors set $U = \{u_1, u_2, \ldots, u_m\}$; it is an indicators set formed by evaluation index;

(2) Reviews set $V = \{v_1, v_2, \ldots, v_n\}$; it is a collection of reviews, and reviews includes excellent, good, medium and bad four grade.

(3) Weight vector $W = (w_1, w_2, \ldots, w_m)$; it is a collection of weight vector, and it is confirmed by 2.2.1 (AHP).

(4) Single factor judgment $\tilde{f}$: $U \rightarrow F(V)$, $u_i \rightarrow \tilde{f}(u_i) = (r_{i1}, r_{i2}, \ldots, r_{in}) \in F(V)$ 。 According to $\tilde{f}$, fuzzy relations can be induced $R \in F(U \times V)$, and $R(u_i, u_j) = \tilde{f}(u_i)(v_j) = r_{ij}$, and that according to R, it can constitute a fuzzy matrix.

$$R = \begin{bmatrix} r_{11} & r_{12} & \cdots & r_{1n} \\ r_{21} & r_{22} & \cdots & r_{2n} \\ \vdots & \vdots & \vdots & \vdots \\ r_{m1} & r_{m2} & \cdots & r_{mn} \end{bmatrix}$$

In factors set U, Weights fuzzy vector $W = (w_1, w_2, \ldots, w_m)$, through transformation

Of into R fuzzy set $B = W \times R$, (U, V, R) in reviews set, it constitutes a comprehensive evaluation model.

3. Empirical research

3.1 Using AHP

Ask a Panel of Experts to have a single factor judgment on the various indicators Enterprise Group, and obtain secondary and tertiary indicators index weights. The results are showed on table 3.1.

3.2 Fuzzy evaluation sets

The various indicators of a certain Enterprise Group measurement

$X = (x_{11}, \cdots, x_{13}, x_{21}, \cdots, x_{22}, \cdots, x_{103})$

$= (80,79,74,92,90,90,92,94,95,76,78,88,92,90,88,76,77,86,80,90,92,72,91,92, 90,77,92,90,76)$,

From the factors set $U = \{u_{11}, \cdots, u_{13}, u_{21}, \cdots, u_{22}, \cdots, u_{103}\}$

Reviews Set $V = (v_1, v_2, v_3, v_4)$ = (excellent, good, medium and bad), obtain the evaluation matrix by Table 3.1.

3.3 Pairs Table 3.1 fuzzy relationship computing in Table 3.2.

The ultimate capacity of the enterprise index for the Synergetic Group:

= X = (0.5111 0.3582 0.1109 0.0510)

Focus grade scores were 30,65,80,95, = (30,65,80,95), which coordinated the ability of enterprise groups integrated into = x = 85.9463.

Table 3.1 Evaluation Matrix and Weight

Level-one Indicator	Level-two Indicator	Level-three Indicator	Excellent	Good	Medium	Bad	Weight (%)
Enterprise Group Synergetic Capacity Evaluation U	Strategic coordination U_1 10.21%	Member enterprises are able to cognize the importance of strategic coordination u_{11}	0.4	0.4	0.2	0	34.28
		Member enterprises are able to coordinate and implement strategic projects u_{12}	0.5	0.3	0.2	0	51.69
		Do complementary strategies to support and share strategic risks u_{13}	0.3	0.4	0.1	0.1	14.03
	Cultural Coordination U_2 9.13%	Group Culture is the carrier of group business philosophy and group awareness, the soup of group u_{21}	0.6	0.3	0.1	0	52.35
		The establishment of common conduct conduce to group coordination u_{22}	0.5	0.4	0.1	0	47.65
	Resources Coordination U_3 15.31%	During stocking, timely deploy and integrate raw materials according to production plans u_{31}	0.5	0.4	0.1	0	22.35
		Production equipment and technical process can timely deploy and integrate according to demand changes u_{32}	0.6	0.3	0.1	0	23.33
		Human resources can timely be adjusted and trained according to the needs of the development of enterprise group u_{33}	0.6	0.4	0	0	26.16
		Founds can be deployed timely according to the company's need u_{34}	0.7	0.3	0	0	28.16

Organizations Coordination U_4 10.11%	The flat organization structure extent u_{41}	0.4	0.4	0.1	0.1	35.28
	The situation of sharing group's public organization u_{42}	0.4	0.5	0.1	0	31.32
	The extent of group diversification u_{43}	0.5	0.3	0.2	0	33.40
System Coordination U_5 9.21%	System takes on powerful implementation u_{51}	0.6	0.3	0.1	0	33.33
	Consistency of all companies' rules and regulations u_{52}	0.5	0.4	0.1	0	33.33
	Rationality and practicability of all regulations u_{53}	0.5	0.3	0.2	0	33.33
Information Coordination U_6 9.12%	Application of information technology in many internet-based companies u_{61}	0.4	0.4	0.1	0.1	40.32
	Enterprise' accumulated knowledge rapidly transfer and circulate in different business units and departments u_{62}	0.4	0.5	0	0.1	28.84
	Focus on information collection, storage, processing, transmission, utilization , and so on u_{63}	0.5	0.3	0.2	0	30.84
Process Coordination U_7 13.72%	Stock together, launch into market of goods supply u_{71}	0.4	0.4	0.2	0	28.32
	Share production capacity, achieve the information exchange of production links with other sectors u_{72}	0.5	0.4	0.1	0	35.84
	Share brand and sales channels, all sides of marketing realize information coordination u_{73}	0.6	0.3	0.1	0	35.84
Innovation Coordination U_8 8.56%	Jointly established R&D organization u_{81}	0.5	0.3	0.1	0.1	33.33
	Communicate innovation experiences and achievements u_{82}	0.6	0.2	0.2	0	33.33
	Share technology and R&D achievement, exchange and use technology each other u_{83}	0.6	0.3	0.1	0	33.33
Leases Coordination U_9 7.01%	Normative acts among members to reduce synergetic risk u_{91}	0.5	0.4	0.1	0	58.87
	Achieve institutional protection and support u_{92}	0.4	0.5	0	0.1	41.13
Commerce Coordination U_{10} 7.62%	Enterprise groups unify internal business language and business platform u_{101}	0.6	0.3	0.1	0	50.17
	Enterprises can conduct business through the internet activities u_{102}	0.5	0.4	0.1	0	25.92

		Achieve commercial data exchange between enterprises u_{103}	0.4	0.4	0.1	0.1	23.91

Table 3.2 Fuzzy Relationship Operational Results

Level-one Indicator	Level-two Indicator	Fuzzy Relationship Operational Results			
Enterprise Group Synergetic Capacity Evaluation U	Strategic Coordination U_1	.4377	.3483	.1860	.0140
	Cultural Coordination U_2	.5524	.3476	.1000	.0000
	Resources Coordination U_3	.6052	.3482	.0456	.0000
	Organizations Coordination U_4	.4334	.3979	.1334	.353
	System Coordination U_5	.5333	.3333	.1333	.0000
	Information Coordination U_6	.4380	.3980	.1020	.0692
	Process Coordination U_7	.5075	.3642	.1283	.0000
	Innovation Coordination U_8	.5666	.2666	.1333	.0333
	Leases Coordination U_9	.4589	.4411	.0589	.0411
	Commerce Coordination U_{10}	.5263	.3498	.1000	.0239

4. Discussion of results

This paper focuses the single enterprise's evaluation model on the synergetic capacity in a Industrial park. That the larger-size group would benefit from that though the factors are not sound, which would be improved in the future research..

5. Conclusion

The paper begins with the concept of enterprise group's synergetic capacity. Then it builds up the model of enterprise group's synergetic capacity by SPSS, and the evaluation system of an enterprise group's synergetic capacity by using level analysis methodology and vague integration evaluation methodology. This method is very feasible and practical. It truly reflects the integrated status of the enterprise group. And it establishes a scientific assessment system and has great significance for strategic decision-making of the enterprise group.

References

[1] Wang Guoshun etc. Enterprises Theory: Capability Theory[M], Beijing, China Economy Press, 2005, (12), 8-18.

[2] Gu Baoguo, Enterprise Group Synergetic Economic Research[D], Fudan University, 2003.

[3] Jing Ranzhe, Enterprise Mechanism of Cluster Systems Development Based on the Self-organization Synergetic Theory [J], Project Management Journal, 2007, 2, ,52-59.

[4] Mao Kening, Du Gang, Enterprise Coordination Capacity and Evaluation Based on Coordination Product Commerce[J], Journal of Inner Mongolia Agricultural University(Social Sciences), 2006, 2, ,166-172.

[5] Pan Kailing, Bai Leihu, Cheng Qi, Systems Thinking of the Multiplier Effect on Coordination Management[J], Systems Science, 2007, 1, , 70-77.

[6] Wu Zhenggang, Han Yuqi, Zhou Yezheng, Study of Enterprise Capability Index Evaluation Model [J] Planning and Management, 2004, 2, , 147-148.

[7] Zhao Tao, Management Common Methods[M], Tianjin University Press, 2006, 7, 316-346.

Research on VEAs-based Environmental Management Models in Eco-industrial Parks

Wu Chunyou[1], Cao Jingshan[1,2]

1 School of Management, Dalian University of Technology, Liaoning, China, 116024

2 China Datang Corporation, Beijing, China, 100032

Abstract: Environmental management based on voluntary environmental agreements (VEAs) is an innovative environmental management instrument footing on industries, voluntary participation and normally taking the form of contracts or agreements to get the environmental quality improved accordingly. In this paper, the major drivers for companies to adopt VEAs have been studied and a utility-maximizing model is set up, which mainly provides theoretical reference and guidance for the operation of eco-industrial parks.

Keywords: Voluntary Environmental Agreements (VEAs), Transaction Cost, Utility-maximizing

1. Introduction

Since 1970s, the control-and-command and market-oriented strategies and measures have been the mainstream policy instruments in the field of environmental management. Control-and-command measures mainly include such regulations as standards for industrial emissions, post-consumption wastes management laws and so on. In practice, the governments, responsible for collecting information concerning implementation in various regions, normally play the role of the coordinating center for a control-and-command policy. Under control-and-command model, the environmental management usually runs at very poor efficiency[7], more reasonable and effective measures are hardly carried out adaptively[2], and the related management costs are over-loaded[3]. On the other hand, market-oriented measures mainly include various kinds of policies such as trade permissions, emissions charges, pollution insurance, individual resource utilization quotas, etc[4]. By means of market mechanisms, the market-oriented measures is taken to effectively reduce the information treatment costs. However,

---

1   Wu Chunyou, professor and doctoral supervisor from Management School of Dalian University of Technology.

2   Cao Jingshan, doctoral candidate from Management School of Dalian University of Technology, director of International Cooperation Department of China Datang Corporation.

compared with that of control-and-command measures, uncertainties in the operations are also increased.

With the view of the sustainable development of industrial parks, the demands for strategic transformation of environmental management made by companies in the industrial parks are becoming more and more stringent recently[1]. Companies are requested to take initiative behavioral strategies[6] to face with the new challenges. Compensating the deficiency of the above-mentioned two environmental management measures, Voluntary Environmental Agreements (VEAs) based on environmental management measures comes into form. After analyzing conceptual meanings and characteristics of VEAs-based environmental management model, this paper investigates four major drives for industries to participate in VEAs-based environmental management. Economically, socially, and morally, a utility-maximizing model with four major drives is set up for decision-making of the industries in terms of VEAs-based environmental management.

2. VEAs: An Innovative Environmental Management Model

VEAs refers to those environmental agreements made by companies/industries and public authorities (governments or non-government organizations), or initiated by public authorities and implemented by companies/industries, or directly made by polluting companies/industries aiming to improve their environmental management performances. Generally, voluntarily agreements are equal to contracts (or obligations). A VEAs-based environmental management model refers to the mutually restricting relationships between governments and companies under voluntary environmental agreements, which aims to improve the environmental management behaviors of companies, enhance environmental qualities and increase resource-utilizing efficiencies. In recent years, VEAs is broadly used in Europe, Japan and America. VEAs provides companies with more freedom to deal with their environmental problems, and proves effectively in the field of the pollution prevention[8].

VEAs-based environmental management model has four major characteristics. Firstly, VEAs will not be executed compulsorily in the form of laws, regulations or other commanding measures. According to institutional economics, VEAs could be grouped into the catalog of informal institutions, which is developed during the long process the people's interactions, not necessarily being implemented by intervention from outside authorities or organizations but through voluntary social interactions. Compared with formal institutions, informal ones normally have longer persisting vital force. VEAs-based environmental management is an embodied form of various social informal institutions such as the social consciousness of resource scarcity, awareness of environmental protection, growing willingness of environmental payoff, and other related ethic and morals etc. The non-compulsory feature of VEAs-based environmental management decides its voluntary characteristics. Secondly, informal institution presents a direction to decrease the transaction costs of formal institutions, or an evolutionary direction from formal institutions to informal institutions. Therefore, the transaction costs of VEAs-based environmental management should be lower than that of other formal environmental management institutions. Thirdly, VEAs-based environmental management is a response made by companies to changes in the environment they are facing with, which leads to an obvious market-economy driving characteristic. Lastly, since agreements are made on a voluntary basis, VEAs-based environmental management is an adaptive and flexible management instrument. Effective implementation of this instrument should be dependent on the support of regulatory systems and also public supervisions, namely being operated under comprehensive regulatory systems including laws contracts and civil regulations on consumers, competitions, health, environment, and so on. Sometimes, VEAs-based environmental management agreements are also a complementary to the above regulations.

Generally, VEAs-based environmental management model has four major characteristics, namely voluntary, low-transaction-cost, market-oriented, and depending on formal regulations and public supervisions. As a flexible VEAs environmental management model, VEAs-based environmental management does not only motivate the voluntary participation of industries, lessen the management burdens of industrial parks, but also help to mollify the possible conflicts between the park and the tenant companies in terms of environmental problems.

3. Drivers for Companies to participate in VEAs

To maximize the utility is the embodiment of companies' social responsibilities in modern societies. To produce virtual social values and to maximize the utility have been generally accepted by economics, and become the compass of companies' survival and development[9,10]. Utility is a comprehensive index. In this paper, we suppose the goal of decisions of companies is to maximize their utilities, and the participation in VEAs is a result of these efforts. Under that pre-condition, and combining perspectives, this paper categorizes the potential drives for companies to participate in VEAs as the matrix shows in figure 1.

The four major drivers will be analyzed in detail as followed.

3.1 Market drivers (external and economic)

Market explanation related to the fundamental goal of companies. In this explanation, drives for companies' participation in VEAs mainly come from the economic motivations in the product and financial markets.

In 1994, Church explained the eco-product certification could solve consumers' trusting problems with the green announcement of industries[11], he furthermore explained the problems of

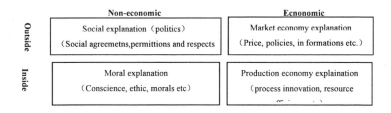

	Non-economic	Ecnonomic

Outside	Social explanation (politics) (Social agreemetns,permittions and respects	Market economy explanation (Price, policies, in formations etc.)
Inside	Moral explanation (Conscience, ethic, morals etc)	Production economy explaination (process innovation, resource

Figure 1   Matrix of drives of companies' participation in VEAs

the transparency in industrial environmental management. According to informative economics, VEAs-based environmental management models can be permitting mechanism in the product market, with which consumers are able to confirm whether the producing process and the final products of some a industry meet the announced environmental requirements or not.

In financial market, it is generally regarded that companies with sound environmental performances will achieve a better financial results. If this hypothesis could be proved, financial investors will take the participation in VEAs as a signal of better financial performance, and invest more in those companies who have taken part in VEAs in order to get more financial rewards. On the other hand, companies participating in VEAs can also get financed at a lower cost and be able to attract those so called "Social responsibility funds".

3.2 Production's economic explanation (internal and economic)

Production's economic explanation means that companies participating in VEAs are motivated by their internal economic factors, which on the contrary will improve the efficiencies of productions and services. By participating in VEAs, the companies would establish their environmental management systems, or innovate the processes and products, which will result in the increase of resource efficiency and performance in turn [12].

3.3 Social explanation (external and non-economic)

Social explanation means that drivers for the companies participation in VEAs depend on the social transaction relationships between their organizations and other stakeholders.   Social transactions include relationships among individuals such as friendships, mutual influences, social assistances, and neighboring friendliness, and those in the organization levels such as competition among organizations, union relationships, etc. Social transactions are "the unclear (non-regulated) obligations of companies"[13]. Whereas, social rewards include social agreements, social acceptances and social respects. The above-mentioned non-economic factors have abstract influences on industries. For example, communities and environmental beneficial groups in which companies are located or connected have an obvious driving effects for companies participate in VEAs, such as changing their pollution behaviors [14], urging them to make environmental protection promises [15], making them to carry out environmental management systems [16] and so on.

Political drive is special aspect of the social explanation. By participating in VEAs, companies will try to obtain a much better environmental performance, such as lower environmental emissions, than regulations and governmental environmental management require. In this case, VEAs will be helpful in terms of coordinating companies' behaviors, reducing their blindness and influencing opinions of stakeholder and policies of the governments.

3.4 Moral explanation (internal and non-economic)

Moral explanation means the reason for a company to participate in VEAs is that it thinks the plan is good. Actually even highest decision-makers of a company will be also influenced by moral and ethic factors. They will also play a crucial role during the formulation of the environmental values and awareness of the company. The important influence of individual awareness is indicated by the companies' participation and may result in an internal moral benefits.

4. Utility-maximization Model: a Political Economic Organization Model

According to the above explanations concerning drivers for companies to participate in VEAs, we can divide the utility-maximizing goals of a company into: (1) current and future interests (according to market and production revenue drives), (2) social admirations to the companies (according to social drives), (3) moral satisfactions (according to moral drives), Therefore, a company utility-maximizing model will be set up as shown by figure 2.

According to above framework, the companies' utilities could be functionalized as the following:

$$U_i = Max \ \{u^1\{\pi \ (X_i) \ , \ s \ (Y_i) \ , \ v \ (Z_i) \ \}, \ u^0\{\pi \ (X_i) \ , \ s \ (Y_i) \ , \ v \ (Z_i) \ \}\}$$

thereas:

u^1 utilities of companies which participate in VEAs;

u^0 utilities of companies which do not participate in VEAs;

$\pi(.)$ interests of the company I, is X_i;

$S(.)$ social admiration of company, customizing variable of its stakeholders is Y_i;

$V(.)$ moral satisfaction of the decision-maker of company;

Z_i customizing variable of individuals or the person investigated.

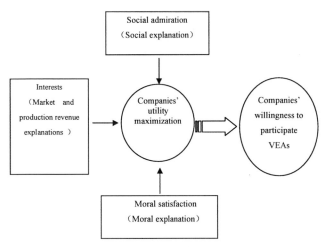

Figure 2. Framework of utility-maximizing model for decision-making

Suppose all the other managing decisions except VEAs-related will be not influenced by the decision in question. Before considering VEAs, all the other management decisions have been made according to company's budgets for the utility-maximizing, the levels of which will not be changed in short terms. Referring to Bale's methodologies in 1974[17], while combining non-economic factors, namely social and moral drives into economic analysis, the individual utility function could be viewed as including individual basic demands and individual characteristics. That means whether the company participates in VEAs or not will be decided by

$\underset{vol}{MAX}, U_i$, that is:

$U_i = u[vol_i/X^*_i \, Y^*_i \, Z^*_i]$

Thereas：

Vol   decision-making variable，vol=1 means "to participate"，vol=0 means "not to participate"；

X*,Y*,Z*   company's characteristics considered while managers make the decisions. All these variables are outside ones while the decision in question is being made.

In contrast to traditional interests-maximizing model, this model take both economic factors and non-economic factors including political ones into consideration, which is helpful to embody the principal features of environmental management. So we also call this model a political economic organization (PEO) model[18].

From the above model, we know that if u[vol=1] > u[vol=0], then the company will participate in VEAs; if u[vol=1]≤u[vol=0], the company will not participate in VEAs.

5. Conclusions and limitations

With the new slogan of pollution prevention, World Bank has taken the initial efforts to combine factors of social interactions into the explanations of the environmental economics since 2000. They take the government, market and community as three major social influential factors. In this paper, those factors are spitted into market, production revenue and social drives. Besides, the moral factor of the decision-maker of the company is also considered through which the analysis scope has been expanded accordingly.

VEAs-based environmental management is innovative for companies. In China, the eco-industrial parks are still in the starting process. The tenant companies are generally featured with small size, underdeveloped technology, mass consumption of natural resources, poor utilization efficiencies, and dispersed mutual connections, etc. The environmental management taken by authorities in the parks has such deficiencies as un-completed regulatory systems, non-restrict executions, inadequate supervisions, low effective motivating mechanisms, etc. Due to the above problems, the tenant companies are passive, in lack of willingness and initiatives. Chinese eco-industrial parks should attach a higher importance and actively promote those motivating and cooperative measures, so as to encourage the tenant companies to achieve better environmental performances over the regulatory requirements in more flexible manners. Thus theoretically implementing VEAs in industrial parks is both feasible and effective.

There are also some limitation with the hypothesis of utility-maximizing goals. Firstly, it is not completely confirmed that whether the decision-makers will make their decisions closely according to this rule or not, or whether the concise calculations would be made during the decision-making process or not. Secondly, this hypothesis also implies the complete rationality of the decision-makers, which is obviously not fact in practices. Therefore, in the future research, both rational and non-rational factors would be considered comprehensively.

References:

1. Tietenberg,T. Emissions trading: An exercise in reforming pollution policy.[Z] Resources for the Future(1985). Washington, D.C.
2. Tietenberg,T. Design lessons from existing air pollution control system:the nited States ,[Z]in S.Hanna and M.Munasinghe, Property rights in a social and ecological context:Case studies and design applications, (1995).Washington D.C. The World Bank.
3. Stanbury&Vertinsky, Governing instruments for forest policy in British Columbia:Apositive and normative analysis.[Z] In C. Tollefson, The wealth of forests: Markets, Regulation, and sustainable forestry,1998 (pp.42-77).Vancouver :University of British Columbia Press.
4. Stavins,R.N. Experience with market-based environmental policy instruments. In K.Maler&J.Vincent(Eds.),[M]The handbook of environmental economics. Amsterdam: North Holland/Elsvier Science(2000).
5. Hawken,P.,Lovins,A.,&Lovins,L.H. Natural Capitalism: Creating the Next Industrial Revolution .[M]Boston,MA:Little, Brown and Company,p8.(1999).
6. Reinhardt, F.LDown to earth. Boston.[M]Harvard Business School Press. (1999).
7. Rie Tsutsumi. The nature of voluntary agreements in Japan, functions of Environment and Pollution Control Agreements. [J] Journal of Cleaner Production 9 (2001) 145-153
8. Tietenberg,TDisclosure. strategies for pollution control.[J]Environmental and Resources Economics.(1998).1193-4,587-602.
9. Levy D. The Environmental practices and Performance of Transnational Corporations.[J]Transnational Corporations,1995,(2): 44-67
10. Becker,GIrrational behavior and economic theory,[J] Journal of Political Economy, .(1962)70,1-13.
11. Church,J.M.A Market solution of green Marketing: Some lessons from the economics of information.[J] Minnesota Law Review, （1994）,79,245-324
12. Porter,M.& van der Linde, C. Green and Competitive: Ending the Stalemate,[J] Harvard Business Review, Sept-Oct 1995:120-134
13. Blau,P.M .Exchange and power in social life. [M]New York: John Wiley, .(1964),91-117.
14. Pargal,S.,&Wheeler,DInformal, regulation of industrial pollution in developing countries: Evidence from Indonesia, [J]Journal of Political Economy, (1996)104,1314-1327
15. Maxwell et al. and C. Decker, Voluntary Environmental Investment and Regulatory Flexibility.[A] Working paper, Department OF Business Economics and Public Policy, Kelly School of Business, Indian University(1998)
16. Florida and Davison, Gaining from green management: Environmental Management systems inside and outside the factory.[J] California Management Review, (2001).43(3):64
17. Becker,G. A theory of social interactions.[J] Journal of Political Economy,82,1063-1093.The essence of Becker. Stanford, California: (1974),.Hoover Institute Press.
18. SoderbaumValues, ideology and politics in ecological economics.[J] Ecological Economics, (1999，2000).28,161-170.

Assessment of Sustainable Development of Eco-industrial Parks based on Material Flow Analysis (MFA)[1]

Shang Hua and Qu Ying

School of Management,Dalian University of Technology, Dalian, P. R. China, 116024

Abstract: Material flow analysis (MFA) has become a useful tool for industrial ecology (IE) to analyze the metabolism of social systems, such as countries and regions. This paper proposes to use the indicators derived from MFA, complemented with water and energy indicators, to analyze the eco-efficiency and the materialization ranks of industrial parks. The methodology is applied to a case study of an industrial park located in Dalian (China). The result shows that new MFA indices are practical in evaluating the Eco-Industrial Park.

Keywords: Eco-industrial parks (EIPs); Material flow analysis (MFA); Sustainable development; Eco-efficiency

0. Introduction

Growing worldwide resource consumption levels, coupled with increasing urbanization and rising population sizes are resulting in increasing pressures on the existing levels of natural resources (Tammemagi, 1999; UNEP, 2000) [1]. A key plank in the calls for greater resource efficiency has been for the incorporation of the concept of sustainability. "Sustainable Development" is a concept that stresses on the balance of economic growth, environmental protection and social equity. Sustainability is now considered to have four dimensions, namely social, economic, environmental and institutional (UNCSD, 2001) [2]. With the greater awareness of the benefits that could be achieved, there is a need for an improved understanding of how such strategies could be developed and implemented. During the stage of economic development, establishment of industrial parks has long been considered as an effective way of achieving the goal of economic growth. Since 1970s industrial parks have been an important part of the economic development strategy in many countries and the number of such parks has grown rapidly throughout

1 Supported by the Initiating Fund of Liaoning Province for Doctors (20081094) & Fund for Philosophy of DUT (DUTHS2008403)

the world. In China industrial parks have emerged rapidly during the last decades in many industrial sectors and the number still keeps on growing[3]. Therefore, quantitative methods to measure and evaluate the performance of the industrial ecology approaches are needed to convince the practitioners of its applicability as a strategic and operational management instrument.

Many people are interested and had made some dedication to the study of performance evaluation of EIP, while most of the studies mainly focus on the setting up of the indicator system by the qualitative methods. And the most puzzles are: how to get the quantitative results? How to compare the different kinds of indicators? How to reflect the whole eco-performance of EIPs ?

1. Defining EIPs

The study of EIPs has assumed great deal of importance within the past ten to fifteen years. Indeed, there is a growing body of literature on the subject (Hawken, 1993; Ehrenfeld and Gertler, 1997; Korhonen et al. 2002; Pellenbarg, 2002; Desrochers, 2002, 2004; Ayres, 2004; Roberts, 2004; Sterr and Ott, 2004; Korhonen and Sankin, 2005)[4]. One of the best definitions of an EIP has been provided by the USEPA (United States Environmental Protection Agency. It states what an EIP is (Martin et al., 1996): A community of manufacturing and service businesses seeking enhanced environmental and economic performance by collaborating in the management of environmental and reuse issues. By working together the community of businesses seeks a collective benefit that is greater than the sum of the individual benefits each company would realize if it optimized its individual performance only[5].

One of the strong points of EIP is the cooperation between businesses to find win-win situations, where the benefits of cooperation should be greater than the sum of individual advantages. Despite the efforts invested and the attractive definition of EIP, there are few experiences and many problems arise when implementing IE in industrial parks [6]. First, the success of EIP depends on many variables, but time and active company participation are crucial to establish networking, especially in the conversion of existing industrial areas. Thus, the project of conversion is usually a slow and progressive evolution towards an EIP. In this context, indicators are necessary and useful in order to objectively reflect and measure this constant evolution. Secondly, the larger the area analyzed, the easier it is to find these types of situations and the greater the opportunities. Nowadays, some interesting projects can be found at regional level [7]. Moreover, this system growth leads to a rise in flows and data analyzed, thus increasing the complexity of the system. This is when the need to find tools to simplify the system emerges. Indicators capable of structuring and simplifying systems data can be one of these tools[8].

2. Methodology

A deficiency of industrial ecology and EID (Ecological industrial development) is the lack of numerical data to support the existing theories. There is an extensive bibliography on indicators and many have been already defined [6]. In general, a good indicator should be credible, transparent, relevant, accurate, measurable, cost effective, comparable, adaptable, able to show changes, and readily understood [10]. Many times the estimation of some indicators requires a high amount of precise data and in other cases complicated ways of estimation result in final numbers with no physical meaning. These are precisely the advantages of material flow analysis (MFA) derived indicators. MFA is a widespread and standardized methodology, for accounting the input and output material flows of a system, and for estimating their derived environmental indicators [9]. All flows and indicators are measured in mass units, giving a physical quantification of the system's material requirements. The methodology has been mostly applied to countries and regions in order to analyze their social metabolism, but not to industrial areas. At this scale it is more frequent to apply input-output analysis, substance flow analysis and other similar methodologies [9], as it is to apply environmental management accounting to companies [10]. However, the added value in applying MFA at this level is quantifying all material flows and finding meaningful and simple indicators that will be able to detect critical points and reflect the state and evolution of the system's metabolism.

MFA is a scientific and quantitative method to set up the criterion of the performance of an industrial park. Specifically, to show the factors that influence the successful development and functioning of an eco-industrial park[8]. The aim of this paper is to adapt MFA methodology to an industrial park and the companies located within it, and to assess the results obtained for its sustainable development. As shown in Fig.1

3. Material Flow Analysis

The Material Flow Accounting method (Schmidt Bleek, 1993; Hinterberger and Stiller, 1998; Bargigli et al., 2004) is aimed at evaluating the environmental disturbance associated with the withdrawal or diversion of material flows from their natural eco-systemic pathways. In this method, appropriate material intensity factors (g/unit) are multiplied by each input, respectively, accounting for the total amount of abiotic matter, water, air and biotic matter that is directly or indirectly required in order to provide that very same input to the system[8]. The resulting material intensities (MIs) of the individual inputs are then separately summed together for each environmental compartment (again: abiotic matter, water, air and biotic matter), and assigned to the system's output as a quantitative measure of its cumulative environmental burden from that compartment (often referred to as "Ecological Rucksack") [9].

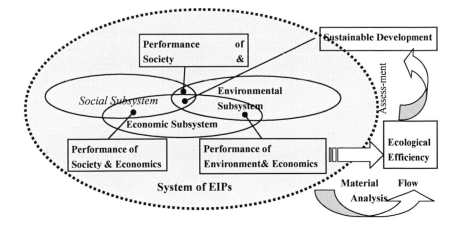

Fig.1 Material flow analysis adapted to an industrial park

Here an attempt is made towards more comprehensive approach using the methods of material flow analysis (MFA), the methodology of which has been systematically developed since the late 1980s (Bringezu, 1993; Bringezu et al., 1995; Hinterberger et al., 2003). Through material flow accounting a meaningful interface between economy and environment is created (Ayres, 1989, 1995, 1996; Erkman, 1997; WCED, 1987), and material flow accounting is being streamlined so as to comply with the structures of the national accounts (EUROSTAT, 2001; SEEA, 2000; UN et al., 2003). The need to unify the concepts and calculation methods has resulted in the handbook for material flow accounting (CEC, 2001). Physical input–output tables have been compiled also in Finland (Mäenpää and Muukkonen, 2001). European Environmental Agency uses the volumes of material flows for continuous monitoring the state of environment and eco-efficiency (Adriaanse et al., 1997; Ayres and Ayres, 1998; EEA, 2000). So far, MFA approach has not been commonly applied specifically to Eco-industrial park sector[10].

4. Analytical framework

4.1. System boundaries

The material flow analysis was applied to each company and to the Eco-industrial parks as a whole. Each company represented a "subsystem" which formed part of the total industrial park, the "system". All subsystems and the overall system were to obey the material flow balance, but the global balance was not the sum of each subsystem's balance. Subsystems boundaries were defined by the territorial limits of the companies and the system area is the sum of the company's areas.

Consequently, if all the companies of the Eco-industrial parks participate in the project, the system boundaries will coincide with the geographical limits of the industrial park. In the case that one company or enterprise has more than one factory, only the factories inside the system boundaries will be included in the analysis.

4.2. Flows

When applying MFA at company level, as well as at national level, indirect flows associated with imports and exports may or may not be accounted. Nevertheless, indirect flows of the whole Eco-industrial park should not be accounted as the sum of the indirect flows associated with each company, in order to avoid double counting. For the same reason, by-products do not have associated indirect flows; except in the case of materials specially used to modify by-product characteristics in order to improve or reuse them. Although water and air are strictly materials, they are considered as independent flows, as in MFA at national level, due to the large amounts used. Material flow indicators only take into account the amount of water and air used in manufacturing processes. The rest is included in independent indicators. As shown in Fig.2.

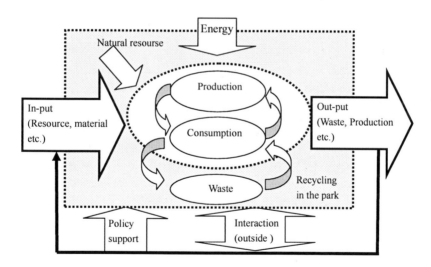

Fig.2 The description of material flows in an industrial park

4.3. Significant issues in the adaptation of MFA to Eco-industrial parks

When MFA for national economies or regions is adapted to Eco-industrial parks, some aspects should be pointed out:

1. MFA should be combined with energy and water flow analysis, as it is important to measure all the resources used by the system. Depending on the park's location, both resources can suppose important environmental impacts. Also, situations in which material decreases at expenses of energy or water consumption (i.e. recycling), should be detected. Finally, in an EIP, as the use of all resources, not only materials, should be improved, all should be quantified.

2. Whereas at national level most of the data are statistical, in Eco-industrial parks data are given by the companies, reflecting better the reality and the system's heterogeneity. Additionally, indirect flows associated to companies' production should be estimated, because the coefficient used at national level could give erroneous values. Furthermore, the results of indirect flows for Eco-industrial parks could be used in the development of statistical basis to estimate indirect flows at national or regional level.

3. In Eco-industrial parks weak dematerialization should be tested measuring material input per unit of product, instead of GDP and per capita which are used at national or regional level.

4. At national level the system is analyzed as a black box, whereas in an industrial park its companies also should be analyzed as well. Thus, in an Eco-industrial park, MFA should be evaluated for the whole system (the whole industrial area) and for its subsystems, in order to measure the companies' flows and to detect opportunities for improvement, such as inter-firm's material exchange.

5. The data basis created to evaluate MFA's indicators will be the main tool to detect these opportunities. Thus, not only the indicators deducted from MFA are useful, but also the data necessary to obtain them.

4.4. Indicators

Two input indicators from the MFA methodology at the national level were used: direct material input (DMI) and total material requirement (TMR). DMI measures the amount of materials entering the system to be used and/or processed, and is the sum of Domestic Extraction (DE) plus Imports. If unused domestic extraction and indirect flows associated with imports are added to DMI, the input indicator is called TMR. Methodologically and conceptually in accordance with MFA indicators, other environmental indicators were deduced and are presented in Table 1. The indicators will be useful to simplify the analysis of the material flows.

Depending on the type of information needed, the indicators could be calculated for the whole area having a global vision of the system, as well as for each company, facilitating the comparison of the consumption or the efficiency of different subsystems. Notwithstanding, material indicators are not enough to analyze the use of resources by industries and their efficiency, and should be complemented with water and energy indicators. The indicators were defined in absolute and in relative terms. Indicators in absolute terms give information about total consumption or generation and can either reflect energy, material or water consumption (inputs to the system), or they can reflect waste production (outputs from the system). On the other hand, indicators expressed in relative terms show the efficiency in the use of materials or water. Considering energy, only consumption indicators were calculated due to the lack of information on energy dissipation in industrial processes. Table 2 shows how MFA indicators can reflect the efficiency of some strategies of conversion of an industrial area into an EIP. In order to form an EIP, a decrease in resources use (material, energy and water) is necessary, as well as indicators which can reflect the change and quantify the efficacy of the strategies implemented.

Table 1 MFA indicators

Indicator	Definition	Expression
TMR (t)	Total-material	Direct material input
DMIw (t/worker)	DMI/worker	DMI/number of workers
TMRw	TMR/worker	TMR/number of workers
TWG (t)	Total-wastes generation	Total amount of wastes produced
TWGw	TWG/worker	TWG/number of workers
WP (t/worker)	Worker productivity	Total production/number of workers
Eco-Ef	Eco-efficiency	Annual production/TMR
Eco-In	Eco-intensity	TMR/total production
M-Inef	Material inefficiency	Outputs to nature/DMI
TWI (t)	Total water input	Total water consumption
TWWG (t)	Total-wastewater	Total amount of waste water produced
TWIw (t/worker)	TWI/worker	TWI/number of workers
TEI (GJ)	Total energy input	Total energy consumption
TEIw (GJ/worker)	TEI/worker	TEI/number of workers
E-In (G/t)	Energetic intensity	TEI/total production

Table 2 Trends of MFA indicators in the implementation of strategies of conversion of industrial areas into EIPs

. By-product exchange	↓ DMI or TMR
	↓ TWG
Strong dematerialization	↓ DMI or TMR
Weak dematerialization	↓ Eco-In
Minimization of outputs to nature	↓ TWG
	↓ M-Inef
Use of system own resources	↓ Imports versus
	↑ Domestic extraction
Strategies to increase the efficiency	
Materials	↓ DMI or TMR
	↓ Eco-In
Energy	↓ TEI or Y E-In
Water	↓ TWI

4.5. Case study

The indicators were calculated for an Eco- industrial park located in Dalian High-Tech Industrial Zone, one of the few special zones in the Peoples Republic of China that are sanctioned by the National Government in March 1991. The Zone, located in an area of approximately 35.6 km^2, has more than 2,300 registered companies. Over 800 of these are international companies. The industrialization growth was fast and unorganized, as in many other parts of the world. More than 30 of these are among the Global Fortune 500. During the past twelve years, the economy of Dalian High-tech Industrial Zone has been growing rapidly. The growth rate is accelerating in recent years especially among companies engaged in software, information service, media and digital entertainment.

4.6 Results

All the indicators were calculated for the whole industrial area and for each company, and the results are presented in Table 3. The table shows the results of each indicator for the "Total system", the Eco-industrial park as a whole; and for all the "subsystems", the average values for the analyzed companies, as well as the highest (maximum) and lowest (minimum) indicator values.

Table 3. Results of MFA indicators for the case study

Indicators	Units	Total	Subsystem's	Subsystem's	Subsystem's
DMI	t/year	223,828	5596	49,751	9
DMIw	t /year/worker	112	183	1418	1.0
TWG	t/year	31,472	787	11,200	0.7
TWGw	t/year/worker	15.8	23	400	0.02
WP	t/year/worker	96.5	160	1412	0.3
Eco-Ef	0.8	0.9	0.8	1.0	0.03
Eco-In	-	-	2.2	29.1	1.0
M-Inef	1.2	0.1	0.2	1.0	0.001
TWI	t/year	357,268	8,932	171,212	61
TWIw	t/year/worker	179	229	4506	6
TWWG	t/year	293,972	7,349	170,000	60.5
TEI	GJ	345,673	8,863	87,537	19.8
TEIw	GJ/worker	174	209	1430	4.0
E-In	G/t	1.8	9.0	68.3	1.6

5. Conclusions

The case study focused on a high-tech industrial area, where individual strategies for selected companies should be combined with collective objectives in order to begin the conversion of this industrial area into an EIP. The indicators showed that one company generates 22% of the system's DMI and seven companies generate 85% of its TWG. Thus, the proposed indicators, by using simple data available for all companies, are useful in detecting the system's critical points: resource consumption (DMI, TMR, TWI, TEI), the usage of system's own resources (domestic versus imported), waste generation (TWG or M-Inef) and efficiency (Eco-Ef or Eco-In).

A high value of an indicator does not represent a critical environmental situation, since the relation between flows and environmental impacts is not direct. Many flows have multiple and complex effects on the environment, so it is difficult to identify and quantify the impacts of each flow. However, it is obvious that the higher the flow, the higher the impact. Therefore a reduction in resource use is needed to make economic development couple with earth carrying capacity. Still one of the advantages is the higher simplicity of data required in applying the methodology proposed, acquainted by all the companies, which facilitates its participation in the project. Despite the fact that many companies should be dropped, the experience proves that the participation would be lower with more data requirements. This participation is crucial for the success of an EIP and, most importantly, the higher the number of participants, the greater the possibilities of interconnections among businesses.

To conclude, it should be remarked that this analysis goes beyond an individual analysis of each company and focuses on the whole Eco-industrial park. The case study, unlike many other cases, has no central company in this industrial area. This paper shows the results for a heterogeneous industrial area, formed by many SMEs from different industrial sectors, similar to many other areas around the world.

References

[1] Cote R. Thinking like an ecosystem. Journal of Industrial Ecology 1998; 2(2): 9-11.
[2] Eurostat. Economy-wide material flow accounts and derived indicators. A methodological guide. Luxembourg: Office for Official Publications of the European Communities; 2001.
[3] Lowe E. Eco-industrial park handbook for Asian developing countries. Oakland, CA: Indigo Development; 2001.
[4] Olsthon X, Tyteca D, Wehrmeyer W, Wagner M. Environmental indicators for business: a review of the literature and standardization methods. Journal of Cleaner Production 2001; 9: 453-463.
[5] Ayres R, Ayres L. Handbook of Industrial ecology. Cheltenham, UK: Edward Elgar; 2002.
[6] Ulgiati, S., Bargigli, S., Raugei, M., Integrated indicators to assess design, performance and environmental sustainability of energy conversion processes. Proceedings of the International Conference Integrative Approaches towards Sustainability, Riga, Latvia. Valero, 2003
[7] Geng, Y., Qinghua, Z. & Keitsch, Eco-Industrial Development in China. 2003.
[8] Cote, R. Designing and Operating Eco-Industrial Parks. 2nd International Conference & Workshop for Eco-Industrial Development, Thailand, 2004 .3.
[9] Brunner P, Rechberger H. Practical handbook of material flow analysis. Boca Raton, FL: CRC Press; 2004.
[10] Bailey R, Allen J, Bras B. Applying ecological input-output flow analysis to material flows in industrial systems. Journal of Industrial Ecology 2004; 8(1-2):45-67.